Lecture Notes
in Business Information Processing 207

More information about this series at http://www.springer.com/series/7911

Andrea Burattin

Process Mining Techniques in Business Environments

Theoretical Aspects, Algorithms, Techniques and Open Challenges in Process Mining

Andrea Burattin
University of Innsbruck
Innsbruck
Austria

ISSN 1865-1348   ISSN 1865-1356   (electronic)
Lecture Notes in Business Information Processing
ISBN 978-3-319-17481-5   ISBN 978-3-319-17482-2   (eBook)
DOI 10.1007/978-3-319-17482-2

Library of Congress Control Number: 2015938082

Springer Cham Heidelberg New York Dordrecht London

Printed on acid-free paper

Springer International Publishing AG Switzerland is part of Springer Science+Business Media
(www.springer.com)

Preface

This book encompasses a revised version of the Ph.D. dissertation, written by the author, at the Mathematics Department of the University of Padua (Italy), and at the Computer Science Department of the University of Bologna (Italy).

In 2014, the dissertation won the "Best Process Mining Dissertation Award", assigned by the IEEE Task Force on Process Mining to the most outstanding Ph.D. thesis, discussed between 2012 and 2013, focused on the area of business process intelligence.

The increasing availability of storage and computing capability, combined with the advent of new "smart" devices, represents the fundamental basis of the so-called "Internet of Things" (IoT). Business companies are focusing their attention to IoT as well, since it could be exploited in a valuable manner. One of the results of such IoT diffusion, but more generally, a common trend of these years, is that the data collection is monumentally increasing.

It is important to remind that the value of data is intimately connected to the *knowledge* that it is possible to synthesize from them. Moreover, in order to strengthen their business, the focus of companies should be on the consolidation and improvement of their business processes, rather than on their data. This is the scenario where process mining sits: in between data mining, and business process modeling.

After a brief presentation of the state of the art of process mining techniques, this book proposes different scenarios for the deployment of process mining projects. In particular, a characterization of companies, in terms of their "process awareness" (and process awareness of their information systems), is detailed.

The work continues identifying and reporting the possible circumstances where problems, both "practical" and "conceptual", can emerge. We identified these three areas as possible sources of problems: *(i)* data preparation (e.g., syntactic translation of data, missing data); *(ii)* the actual mining phase (e.g., mining algorithm exploiting all data available); and *(iii)* results interpretation. Several problems are not limited to a single phase, but orthogonal to all the mentioned sources: for example, the configuration of parameters by non-expert users or the computational complexity of some techniques. In this book we will analyze at least one solution for each of the presented problems. The descriptions of these solutions are kept general, in order to easily allow their tailoring into specific application domains.

The solutions proposed in this book belong to two different computational paradigms: the first considers the classical "batch process mining" (also known as "off-line"); the second introduces the "on-line process mining".

Concerning batch process mining, we are going to investigate first the data preparation problem and we will analyze and present a solution for the problem of hidden data (i.e., when a required field is not explicitly indicated). In our example we are going to consider the "case-id". In particular, our approach tries to identify this missing information by looking at *metadata* recorded for each event.

After that, we will concentrate on the second step (the mining phase) and, in particular, on the problem of exploiting all the available information. As example, we propose the generalization of a well-known control-flow discovery algorithm (i.e., Heuristics Miner) in order to exploit non-instantaneous events. The usage of interval-based recording leads to an important improvement of the algorithm performance. As another example of data exploitation, we present an automatic approach for the extension of a control-flow model with social information (i.e., roles), in order to simplify the analysis of these two perspectives (the control-flow and resources) combined.

Later on, we will focus our attention on another important and, for non-expert users, impacting problem: the parameters configuration. As example, we considered the configuration of a control-flow discovery algorithm. Our approach consists of two steps: first, we introduce a method to automatically discretize the space of parameter values. Then, we present two approaches to select the "best" parameters configuration. The first, completely autonomous, uses the Minimum Description Length principle to balance the model complexity and the data explanation; the second requires human interaction to navigate a hierarchy of models and find the most suitable result.

The data interpretation and results evaluation phase is not problem free, as well. Also in this case, we will analyze the problems and propose two new metrics: a *model-to-model* and a *model-to-log* (the latter considers models expressed in declarative language).

The final part of this book deals with the adaptation of process mining to on-line settings. We will consider, as example, the problem of on-line control-flow discovery. Specifically, we are going to propose a formal definition of the problem and then present two baseline approaches. These two basic approaches are used only for validation purposes. The actual mining algorithms proposed will be two: the first is the adaptation, to the control-flow discovery problem, of a well-known frequency counting algorithm (i.e., Lossy Counting); the second constitutes a framework of models which can be used for different kinds of streams (for example, stationary streams or streams with concept drifts)

Innsbruck, Austria   Andrea Burattin
February 2015

Acknowledgments

I would like to thank, *in primis*, my Ph.D. supervisor: Alessandro Sperduti. His continuous, expert, and passionate guidance incredibly simplified my job. It is a privilege to work with such a generous person and qualified professor and researcher.

I want to express my authentic gratitude to Roberto Pinelli, from Siav. He has been always willing to help me, by all means, and many parts of this book are due to the opportunities he gave me.

Also, I'm very thankful to Paolo Baldan, Diogo Ferreira, Tullio Vardanega, and Barbara Weber who spent their time reading my Ph.D. thesis, and sharing their useful comments.

As mentioned, this book comes as an elaborated version of my Ph.D. thesis. I'm particularly thankful to the organizers of the *Best Process Mining Dissertation Award*: Dirk Fahland, Antonella Guzzo, and Marcello La Rosa. Their detailed comments and elaborate suggestions substantially helped me in shaping this work.

Special thanks go to Wil van der Aalst: working with him and his team has been an incredibly formative experience. His remarkable professionalism and competence are sources of inspiration for my work.

I would like to thank all my colleagues and friends, who shared with me the Ph.D. journey, at the University of Padua and Bologna, in Siav, and at the AIS group, in Eindhoven.

Infine, ringrazio mia moglie Serena, i miei genitori Stefania e Antonio, e tutta la mia famiglia per non avere mai lesinato nel darmi aiuto, fiducia e serenità, e la possibilità di raggiungere i miei obiettivi.

Innsbruck, Austria   Andrea Burattin
February 2015

Contents

Chapter 1
Introduction

For some years now, the usage of information systems has been rapidly growing, in companies of all kinds and sizes. New systems are moving from supporting single functionalities towards a business processes orientation. In Computer Science, a new research area is emerging, called "process mining", which provides algorithms, techniques and tools to perform fact-based analyses, with the final aim of improving business processes.

1.1 Business Process Modeling

Activities that companies are required to perform, in order to complete their own business, are becoming more complex and need the interaction of several persons and heterogeneous systems. A possible approach to simplify the management of the business is based on the division of operations into smaller "entities" and on the definition of the required interactions among them. The term "business process" refers to this set of activities and interactions.

A simplification of a business process, that describes the handling of an order submitted through an e-commerce website, is depicted in Fig. 1.1. In this case, the process is represented just as a dependency graph: each box represents an activity, and connections between boxes indicate the order in which activities may be executed. Specifically, in the example of the figure, the process starts with the registration of the order and the registration of the payment. Once the payment registration is complete, two activities may execute concurrently (i.e. there is no dependency between them). Finally, when the "Goods wrapping" and "Shipping note preparation" are complete, the final "Shipping" activity can start. The conclusion of this last activity terminates the current process instance too.

Most of the software used to define and to help operators involved in executing such processes, typically, leave a trace of the performed activities. An example of such trace (called "log") is presented in Table 1.1. As can be observed, the fundamental information we are going to need to accomplish the mission of this book are the name

© Springer International Publishing Switzerland 2015

A. Burattin: *Process Mining Techniques in Business Environments*, LNBIP 207,
DOI 10.1007/978-3-319-17482-2_1

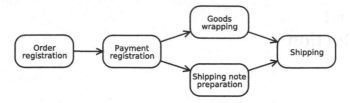

Fig. 1.1 Example of a process model which describes a general process of order management, starting from its registration to the shipping of the goods.

of the activity and the time the activity is executed; moreover, it is important to note that the traces are grouped into "instances" (or "cases"): typically, it is necessary to handle several orders at the same time, and therefore the process is required to be concurrently instantiated several times as well. These instances are identified by a "case identifier" (or "instance id"), which is another field typically included in the log of the traces.

Companies, especially small and medium ones, do not often perform their work according to a formal and explicit business process; instead, typically, they execute their activities with respect to an implicit sorting. Even if such a prescriptive model is not available, the systems used to execute the process, frequently writes executions of process steps into a log file. So, the key idea is that a log can exist even if it is not an expression of a explicit process model (as shown in Fig. 1.1). The aim of process mining is to use such logs to extract a business process model coherent with the recorded events. This model can then be used to improve the company business by detecting and solving deadlocks, bottlenecks, ...

Table 1.1 An example of log recorded after two executions of the business process described in Fig. 1.1.

#	Activities	Execution Time
	Instance 1	
1	Order registration	Feb 21, 2015 12:00
2	Payment registration	Feb 22, 2015 09:00
3	Goods wrapping	Feb 26, 2015 08:30
4	Shipping note preparation	Feb 26, 2015 09:30
5	Shipping	Feb 26, 2015 10:15
	Instance 2	
1	Order registration	Feb 23, 2015 15:45
2	Payment registration	Feb 25, 2015 17:31
3	Shipping note preparation	Feb 26, 2015 08:30
4	Goods wrapping	Feb 26, 2015 10:00
5	Shipping	Feb 26, 2015 12:30

1.2 Process Mining

The term "process mining" refers to an emerging research field which deals with several different activities, all joined by the final aim of extracting knowledge out of the available log files. These activities are: *control-flow discovery*; *conformance checking*; and extension or *enhancement*. We will provide details for each of these activities in Chap. 5 but, in this section, we are going to consider the control-flow discovery as an illustrative example to demonstrate how logs can be treated.

An ideal process mining algorithm for control-flow discovery, as it analyzes the log, identifies all the process instances, and then tries to define some relations among activities. Considering the example of Table 1.1, "Order registration" is always the first activity executed; it is always followed by "Payment registration" and this could mean that there is a *causal dependency* between them (i.e. "Order registration" is required by "Payment registration"). The algorithm continues and detects that "Payment registration" is sometimes followed by "Goods wrapping" and other times by "Shipping note preparation" but, in any case, both activities are always performed. A possible interpretation of such behaviour is that there is no specific order between the execution of the last two activities (which can be executed concurrently), but both of them require "Payment registration". At the end, "Shipping" is observed as last activity always executed after "Shipping note preparation" or "Goods wrapping". Once all these relations are available, it is possible to combine them in order to construct the mined model. The algorithm just presented as example is, essentially, the Alpha algorithm [162] that will be described in Sect. 5.1.

The procedure presented in the previous paragraph helps us to illustrate the general idea of process mining as control-flow discovery: many other algorithms have been designed and implemented, using different approaches and starting from different assumptions. A detailed review of several algorithms will be later provided in this book, in Chap. 5.

However, even if several approaches are available, many important problems remain still unresolved. Some of them are presented in [166], and here we report the most important ones:

- some process models may have the same activity appearing several times, in different positions. However, almost all process mining techniques are not able to extract this kind of tasks: instead, they just insert one activity in the mined model, and therefore the connections of the mined model are very likely to be wrong;
- many times, logs report a lot of data not used by mining algorithms (e.g., detailed timing information, such as distinguishing the starting from the finishing time of an event). This information, however, can be used, by mining algorithms, to improve the accuracy of mined models;
- current mining algorithms do not perform an "holistic mining" of different perspectives, coming from different sources: for example, not only the control-flow, but also a social network with the interactions between the activity originators (creating a global process description). Such global perspective is able to give many more insights, with respect to the single perspectives;

- dealing with noise and incompleteness: "noise" identifies an uncommon behaviour, that should not be described in the mined model; "incompleteness" represents the lack of some information required for performing the mining task. Almost all business logs are affected by these two problems, and process mining algorithms are not always able to properly deal with them;
- visualization of mining results: present the results of process mining in a way that allows people to gain insights in the process.

The key point is that, even if some algorithms solve a subset of the problems, some of them are not solved yet, or the proposed solutions are not always feasible. In this book we try to tackle some of these problems, in order to provide viable solutions for applying process mining in practice.

In this book we are going to analyze the applicability of process mining in real-world business environments. Specifically, when these techniques are applied in reality some of the problems listed previously become evident and new problems (not strictly related to process mining) can emerge. The most outstanding ones are:

P-01 Incompleteness: obtaining a complete log where all the required information are actually available (e.g. in some applications the case identifier might be missing). A log which does not contain all required information, is not useful;

P-02 Exploiting as much information, recorded into log files, as possible (such as resource, time, data), as previously mentioned;

P-03 Difficulties in using process mining tools and configuring algorithms. Typical process mining users are non-expert users, therefore it is hard for them to properly configure all the required parameters;

P-04 Results interpretation: generation of the results with an as-readable-as-possible graphical representation of the process, where all the extracted information are represented in a simple and understandable manner. Non-expert users may have no specific knowledge in process modeling;

P-05 Computational power and storage capacity required: small and medium sized companies may not be able to cope with the technological requirement of large process mining projects.

Our contribution is to analyze the above-mentioned problems, and to propose some feasible directions towards their resolution.

1.3 Book Outline

The structure of the main contributions of this book (i.e., Parts III and IV) are graphically reported in Fig. 1.2. In particular, we are going to analyze all the problems that might occur during a business process mining project. Black texts indicate the objects we deal with in this context; red texts represent chapters of this book; finally, arrows are used to present connections between them. Other chapters are not included in the picture for readability purposes.

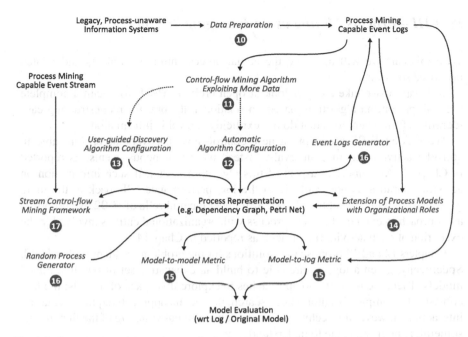

Fig. 1.2 Organization of the main contributions of the book. Each activity is written in italic red font and, the white number in the red circle indicates the corresponding chapter number. Dotted lines indicate that the input/output is not an "object" for the final user, instead it represents a methodological approach (e.g., a way to configure parameters).

Part I – State of the Art: BPM, Data Mining and Process Mining

In this part we are going to provide the fundamental preliminaries and the basic notions (such as a business process, Petri net, BPMN, clustering, metrics), required in order to understand the rest of the book. Some state-of-the-art summary on business process management (Chap. 2), data extraction (Chap. 3), data mining (Chap. 4), and process mining (Chap. 5) is provided as well. This part terminates with some notes on the quality criteria useful in process mining, reported in Chap. 6, and some notions on event streams, described in Chap. 7.

Part II – Obstacles to Process Mining in Practice

This short part will describe, in Chap. 8, the typical application scenarios for process mining project, and the common problems that analysts may have to face. A general long-term view scenario, for process mining projects, is proposed as well in Chap. 9.

Part III – Process Mining as an Emerging Technology

With this part we will dive into the actual process mining problems and related proposed solutions.

In Chap. 10 we take a closer look at open problem **P-01**, to obtain a complete log. We present an algorithm, based on relational algebra, for reconstructing case identifiers from event logs that do not explicitly carry this information.

We tackled **P-02** proposing a new control-flow discovery algorithm, able to consider activities as time intervals, instead of instantaneous events, as reported in Chap. 11. An important aspect of this algorithm is that, if such information on activities duration is not available in the log, performance falls back to the more general case, with no additional cost. Still under the umbrella of **P-02**, we proposed an approach to extend a business process with organizational entities involved in the execution of each activity (i.e., roles), as reported in Chap. 14.

Chapters 12 and 13 propose two solutions for the problem we identified as **P-03**. Specifically, given a log, we are able to build an exhaustive set of possible mined models. Then, we present two approaches to explore this space of models: the first consists of a completely autonomous search; the second approach requires the user's interaction, however this technique is based on the actual *structure* of the final mode: something the user is able to understand.

Concerning the interpretation of results, referred as **P-04**, a model-to-model metric is proposed. This metric, specifically designed for process mining tasks, is able to discriminate business models and can be used for the clustering of processes. In this book, a model-to-log metric is proposed as well. Such metric can give healthiness measures of a declarative process model with respect to a particular log (i.e., if the behavior observed in the log is consistent with the given model). Both approaches are discussed in Chap. 15.

Additionally, Chap. 16 reports problems related to the lack of test data and the approach we built for the random generation of business processes and logs.

Part IV – A New Challenge in Process Mining

Finally, since many times a "batch" approach is not feasible, to address **P-05**, a completely new approach is proposed. We refer to this problem as online process mining (i.e., process mining applied to *event streams*). This new class of techniques allows the incremental mining of streams of events. The approach, which can be used in online manner and is also able to cope with concept drifts, is described in Chap. 17.

Part V – Conclusions and Future Work

This part, with Chap. 18, concludes the book by wrapping-up the content and proposing some possible way to continue the work on these topics.

1.4 Website

All the software described in this book, the datasets, and other additional tools can be downloaded from the book website:

http://andrea.burattin.net/monograph

On this website it is also possible to find further information and the book *errata*.

Part I
State of the Art: BPM, Data Mining and Process Mining

This part gives a general introduction to the process mining field, starting from the very basic notion of business process.

A detailed presentation of the state of the art of process mining is offered, focusing on Control-Flow Discovery algorithms, and some approaches for the evaluation of business processes are described. Data mining, event streams, and quality measures are reported as well.

Chapter 2
Introduction to Business Processes, BPM, and BPM Systems

This chapter provides a basic overview on business processes. In particular it concentrates on the actual definition and characterization of processes, and on the different languages that can be used to describe them. Some notions on the main components of BPM systems conclude this chapter.

2.1 Introduction to Business Processes

It is very common, in industrial settings, that the performed activities are repetitive and have several persons involved. In these cases, it is very useful to define a standard procedure that everyone can follow. A *business process*, essentially, is the definition of such "standard procedure".

Since the process aims at standardizing and optimizing the activities of the company, it is important to keep the process up to date and as flexible as possible, in order to meet the market requirements and the business objectives.

Business Process

There are several definitions of "business process". The most influential ones are reported in [91]. The first, presented in [76] by Hammer and Champy, states that a business process is:

> A collection of activities that takes one or more kinds of input and creates an output that is of value to the customer. A business process has a goal and is affected by events occurring in the external world or in other processes.

In another work, by Davenport [43], a business process is defined as:

> A structured, measured set of activities designed to produce a specified output for a particular customer or market. [...] A process is thus a specific ordering of work activities across time and place, with a beginning, an end, and clearly identified inputs and outputs: a structure for action.

© Springer International Publishing Switzerland 2015
A. Burattin: *Process Mining Techniques in Business Environments*, LNBIP 207,
DOI 10.1007/978-3-319-17482-2_2

In both cases, the main focus is on the "output" of the actions that must take place. The problem is that there is no mention to the originators (i.e., the executors) of such activities and how they are interoperating.

In [113], a business process is viewed as something that: (*a*) contains purposeful activities; (*b*) is carried out, collaboratively, by a group of humans and/or machines; (*c*) often crosses functional boundaries; (*d*) is invariably driven by the outside world. van der Aalst, Weijters and Medeiros, in [167], gave attention to the originators of the activities:

> By process we mean the way an organization arranges their work and resources, for instance the order in which tasks are performed and which group of people are allowed to perform specific tasks.

Ko, in his "*A Computer Scientist's Introductory Guide to Business Process Management*" [91], gave his own definition of business process:

> A series or network of value-added activities, performed by their relevant roles or collaborators, to purposefully achieve the common business goal.

A formal definition of business process is presented by Agrawal, Gunopulos and Leymann in [3]:

> A business process P is defined as a set of activities $V_P = \{V_1, \ldots, V_n\}$, a directed graph $G_P = (V_P, E_P)$, an output function $o_P\colon V_P \to \mathbb{N}^k$ and $\forall (u, v) \in E_P$ a boolean function $f_{(u,v)} = \mathbb{N}^k \to \{0, 1\}$.

In this case, the process is constructed in the following way: for every completed activity u, the value $o_P(u)$ is calculated and then, for every other activity v, if $f_{(u,v)}(o_P(u))$ is "true", v can be executed. Of course, such definition of business process is hard to be handled by business people, but is useful for formal modeling purposes.

More general definitions are given by standards and manuals. For example, the glossary of the BPMN manual [112] describes a process as "*any activity performed within a company or organization*". The ISO 9000 [125] presents a process as:

> A set of activities that are interrelated or that interact with one another. Processes use resources to transform inputs into outputs. Processes are interconnected because the output from one process becomes the input for another process. In effect, processes are "glued" together by means of such input output relationships.

Right now, no general consensus has been reached on a specific definition. This lack is due to the size of the field and to the different aspects that every definition aims to point out.

In the context of this work, it is not important to fix one definition: each definition highlights some aspects of the global idea of business process. The most important issues, that should be covered by a definition of business process, are:

1. there is a finite set of **activities** (or **tasks**) and their executions are partially ordered (it's important to note that not all the activities are mandatory in all the process executions);
2. each activity is executed by one or more **originators** (can be humans or machines or both);

3. the execution of every activity produces some **output** (as a general notion, with no specific requirement: it can be a document, a service or just a "flag of state" set to "executed") that can be used by the following activity.

This is not intended to be another new definition of business process, but it's just a list of the most important issues that emerge from the definitions reported above.

Representation of Business Processes

Closely related to Business Processes is Business Process Management (BPM). van der Aalst, ter Hofstede and Weske, in [161], define BPM as:

Supporting business processes using methods, techniques, and software to design, enact, control, and analyze operational processes involving humans, organizations, applications, documents and other sources of information.

From this definition, it clearly emerges that two of the most important aspects of BPM are **design** and **documentation**. The importance of these two tasks is clear if one thinks about the need to communicate some specific information on the process that has been modeled. The main benefits of adopting a clear business model can summarized in the following list:

• it is possible to increase the visibility of the activities, that allows the identification of problems (e.g. bottlenecks) but also areas of potential optimization and improvement;
• grouping the activities in "department" and grouping the persons in "roles", in order to better define duties, auditing and assessment activities.

For the reasons just explained, some characteristics of a process model can be identified. The most important one is that a model should be **unambiguous**, in the sense that the process is precisely described without leaving uncertainties to the potential reader.

There are many languages that allow the modeling of systems and business processes. The most used formalisms for the specification of business processes have in common to be graph-based representations, so that nodes, typically, represent the process tasks (or, in some notations, also the states and the possible events of the process); arcs represent ordering relations between tasks (for example, an arc from node n_1 to n_2 represents a dependency in the execution so that n_2 is executed only after n_1). Two of the most important graph based languages are: Petri nets [111, 118, 146, 188] and BPMN [112][1].

---

[1]Another language for the definition of "processes" is, for example, the Π-calculus [115, 185]: a mathematical framework for the definition of processes whose connections vary based on the interaction. Actually, it is not used in business contexts and by non-expert users because of its complexity. With other similar languages (such as Calculus of Communicating Systems, CCS and Communicating Sequential Processes, CSP) the situation is similar: in general, mathematical approaches are suitable for the definition of interaction protocols or for the analysis of procedures (such as deadlock identification) but not for business people.

2.1.1 Petri Nets

Petri nets, proposed in 1962 in the Ph.D. thesis of Carl Adam Petri [119], constitute a graphical language for the representation of a process. In particular, a Petri net is a bipartite graph, where two types of nodes can be defined: transitions and places. Typically, transitions represent activities that can be executed, and places represent states (intermediate or final) that the process can reach. Edges, always directed, must connect a place and a transition, so an edge is not allowed to connect two places or two transitions. Each place can contain a certain number of tokens and the distribution of the tokens on the network is called "marking". In Fig. 2.1 a small Petri net is shown; circles represent places, squares represent transitions.

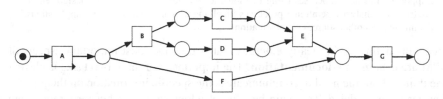

Fig. 2.1 Petri net example, where some basic patterns can be observed: the "AND-split" (activity B), the "AND-join" (activity E), the "OR-split" (activity A) and the "OR-join" (activity G).

Petri nets have been studied in depth from many points of view, such as from their clear semantic to a certain number of possible extensions (such as time, color, …). A formal definition of Petri net, as presented, for example, in [145], is the following:

Definition 2.1 (Petri net). A Petri net is a tuple (P, T, F) where: P is a finite set of places; T is a finite set of transitions, such that $P \cap T = \emptyset$, and $F \subset (P \times T) \cup (T \times P)$ is a set of directed arcs, called flow relation.

The "dynamic semantic" of a Petri net is based on the "firing rule": a transition can *fire* (i.e., be executed) if all its "input places" (places with edges entering into the transition) contain at least one token. The firing of a transition generates one token for all its "output places" (places with edges exiting from the transition). The distribution of tokens among the places of a net, at a certain time, is called "marking". With this semantic, it is possible to model many different behaviors, for example, in Fig. 2.2, three basic templates are proposed. The sequence template describes the causal dependency between two activities (in the figure example, activity B requires the execution of A); the AND template represents the concurrent branching of two or more flows (in the figure example, once A is terminated, B and C can start, in no specific order and concurrently); the XOR template defines the mutual exclusion of two or more flows (in the figure example, once A is terminated, only B or C can start). Figure 2.3 proposes the same process of Fig. 2.2 with a different marking (after the execution of activities A, B and C).

An important subclass of Petri nets is the Workflow nets (WF-net), whose most important characteristic is to have a dedicated "start" and "end":

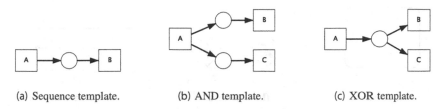

(a) Sequence template.   (b) AND template.   (c) XOR template.

Fig. 2.2 Some basic workflow templates that can be modeled using Petri net notation.

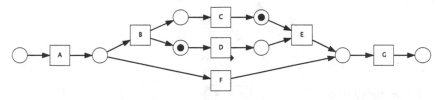

Fig. 2.3 The marked Petri net of Fig. 2.1, after the execution of activities A, B and C. The only enabled transition, at this stage, is D.

Definition 2.2 (WF-net). A WF-net is a Petri net $N = (P, T, F)$ such that:

a. P contains a place i with no incoming arcs (the starting point of the process);
b. P contains a place o with no outgoing arcs (the end point of the process);
c. if we consider $\bar{t} \notin P \cup T$, and we use it to connect o and i (so to obtain the so called "short-circuited" net: $\overline{N} = (P, T \cup \{\bar{t}\}), F \cup \{(o, \bar{t}), (\bar{t}, i)\})$, the new net is strongly connected (i.e. there is a direct path between any pair of nodes).

2.1.2 BPMN

BPMN (Business Process Modeling and Notation) [112] is the result of an agreement among multiple tool vendors, that agreed on the standardization of a single notation. For this reason, now it is used in many real cases and many tools adopt it daily. BPMN provides a graphical notation to describe business processes, which is, at the same time, intuitive and powerful (it is able to represent complex process structure). It is possible to map a BPMN diagram to an execution language, BPEL (Business Process Execution Language).

The main components of a BPMN diagram, presented in Fig. 2.4, are:

Events: defined as *"something that "happens" during the course of a process"*; typically they have a cause (trigger) and an impact (result). Each event is represented with a circle (containing an icon, to specify some details), as in Fig. 2.4(d). There are three types of events: *start* (single narrow border), *intermediate* (single thick border) and *end* (double narrow border).
Activities: this is the generic term that identifies the work done by a company. In the graphical representation they are identified as rounded rectangles. There are few

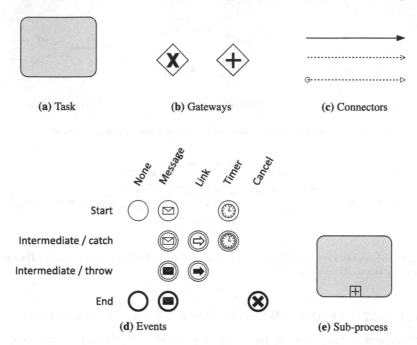

Fig. 2.4 Example of some basic components, used to model a business process using BPMN notation.

types of activity like tasks (a single unit of work, Fig. 2.4(a)) and subprocesses (used to hide different levels of abstraction of the work, Fig. 2.4(e)).

Gateway: structure used to control the divergences and convergences of the flow of the process (fork, merge and join). An internal marker identifies the type of gateway, like "exclusive" (Fig. 2.4(b), on the left), "event based", "inclusive", "complex" and "parallel" (Fig. 2.4(b), on the right).

Sequence and Message Flows and Associations: connectors between components of the graph. A sequence flow (Fig. 2.4(c), top) is used to indicate the order of the activities. Message flow (Fig. 2.4(c), bottom) shows the flow of the messages (as they are prepared, sent and received) between participants. Associations (Fig. 2.4(c), middle) are used to connect artifacts with other elements of the graph.

Beyond the components just described, there are also other entities that can appear in a BPMN diagram, such as artifacts (e.g. annotations, data objects) and swimlanes.

Figure 2.5 proposes a simple process fragment. It starts on Friday, executes two activities (in the figure, "Receive Issue List" and then "Review Issue List") and then checks if a condition is satisfied ("Any issues ready"); if this is the case, a discussion can take place a certain number of times ("Discussion Cycle" sub process), otherwise the process is terminated (and the "End event" is reached, marked as a circle with the bold border). There are, moreover, intermediate events (marked with the double border): the one named A is a "throw event" (if it is fired, the flow continues to the

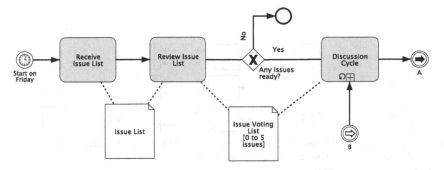

Fig. 2.5 A simple process fragment, expressed as a BPMN diagram. Compared to a Petri net (as in Fig. 2.1), it contains more information and details but it is more ambiguous.

intermediate catch event, named *A*, somewhere in the process but not represented in this figure); the *B* is a "catch event" (it waits until a throw events fires its execution).

2.1.3 YAWL

YAWL (Yet Another Workflow Language) [159] is a workflow language born from a rigorous analysis of the existing workflow patterns [160].

The starting point for the design of this language is the identification of the differences between many languages: out of this, authors collected a complete set of workflow patterns. This set of possible behaviors inspired authors to develop YAWL, which starts from Petri net and adds some mechanisms to allow a *"more direct and intuitive support of the workflow patterns identified"* [160]. However, as authors stated, YAWL is not a "macro" package on top of high-level Petri nets: it is possible to map a YAWL model to any other Turing complete language.

Figure 2.6 presents the main components of a YAWL process. The main components of a YAWL model are:

Task: represents an activity, as in Fig. 2.6(a). It is possible to execute multiple instances of the same task at the same time (so to have many instances of the process running in parallel). Composite tasks are used to define hierarchical structure: a composite task is a container of another YAWL model.

Conditions: a condition Fig. 2.6(b) in YAWL has the same meaning of "place" for Petri nets (i.e. the current state of the process). There are two special conditions, i.e., "start" (with a triangle inscribed) and "end" (with a square inscribed), like for WF-nets (Definition 2.2).

Splits and Joins: a task can have a particular split/join semantic. In particular, it is possible to have tasks with an AND (whose behavior is the same of the Petri

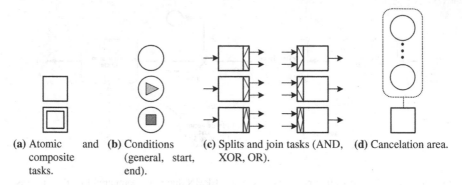

(a) Atomic   and **(b)** Conditions   **(c)** Splits and join tasks (AND,   **(d)** Cancelation area.
     composite   (general,   start,   XOR, OR).
     tasks.   end).

Fig. 2.6 Main components of a business process modeled in YAWL.

net case, presented in Fig. 2.2(b)), XOR (same as Petri net, Fig. 2.2(c)) or OR
semantic. In the last case one or more outgoing arcs are executed[2].

Cancellation Areas: all the tokens in elements within a cancellation area (the dotted
   area in Fig. 2.6(d)), are removed after the activation of the corresponding task
   (whose enabling does not depend on the tokens on the cancellation area).

2.1.4 Declare

Imperative process modeling languages such as BPMN, Petri nets, etc., are very
useful in environments that are stable and where the decision procedures can be
predefined. Participants can be guided according to such process models. However,
they are less appropriate for environments that are more variable and that require
more flexibility. Consider, for instance, a doctor in a hospital dealing with a variety
of patients that need to be handled in a flexible manner. Nevertheless, some general
regulations and guidelines can be followed. In such cases, *declarative* process models are more effective than the imperative ones [122, 156, 187]. Instead of explicitly
specifying all possible sequences of activities in a process, declarative models implicitly define the allowed behavior of the process with constraints, i.e., rules that must be
followed during execution. In comparison to imperative approaches, which produce
"closed" models (what is not explicitly specified is forbidden), declarative languages
are "open" (everything that is not forbidden is allowed). In this way, models offer
flexibility and at the same time remain compact.

   While in imperative languages designers tend to forget incorporating some possible scenarios (e.g., related to exception handling), in declarative languages, designers tend to forget certain constraints. This leads to underspecification rather than
overspecification, i.e., people are expected to act responsibly and are free to select
scenarios that may seem out-of-the-ordinary at first sight.

---

[2]In the case of OR-join, the semantic is a bit more complex: the system needs only one input token,
however if more then one token is coming, the OR-join synchronizes (i.e. waits) them.

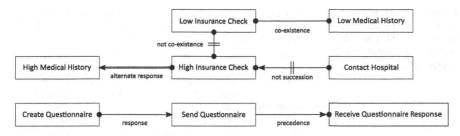

Fig. 2.7 Declare model consisting of six constraints and eight activities.

Figure 2.7 shows a simple Declare model [120, 121] with some illustrative constraints for an insurance claim process. The model includes eight activities (depicted as rectangles, e.g., *Create Questionnaire*) and six constraints (shown as connectors between the activities, e.g., *not co-existence*). The *not co-existence* constraint indicates that *Low Insurance Check* and *High Insurance Check* can never coexist in the same trace. On the other hand, the *co-existence* constraint indicates that if *Low Insurance Check* and *Low Medical History* occur in a trace, they always co-exist. If *High Medical History* is executed, *High Insurance Check* is eventually executed without other occurrences of *High Medical History* in between. This is specified by the *alternate response* constraint. Moreover, the *not succession* constraint means that *Contact Hospital* cannot be followed by *High Insurance Check*. The *precedence* constraint indicates that, if *Receive Questionnaire Response* is executed, *Send Questionnaire* must be executed before (but if *Send Questionnaire* is executed this is not necessarily followed by *Receive Questionnaire Response*). Finally, if *Create Questionnaire* is executed, it is eventually followed by *Send Questionnaire* as indicated by the *response* constraint.

More details on the Declare language will be provided in Subsect. 15.2.1.

2.1.5 Other Formalisms

The language for the definition of business processes, briefly presented in the previous sections, are only a very small fragment of all the available ones. In Table 2.1 some standards are proposed, with their background (either academic or industrial), the type of notation they adopt, if they are standardized somehow, and their current status.

2.2 Business Process Management Systems

It is interesting to distinguish, from a technological point of view, business process design and business process modeling: the first refers to the overall process design (and all its activities); the latter refers to the actual way of representing the process (from a "language" point of view).

In the Gartner's position document [80], a software is defined "BPM-enabled" if allows to work on three parts: integration, runtime environment and rule engine.

Table 2.1 Extraction from Table 2 of [91] where some prominent BPM standards, languages, notations and theories are classified.

Language	Background	Notation	Standardized	Current status
BPDM	Industry	Interchange	Yes	Unfinished
BPEL	Industry	Execution	Yes	Popular
BPML	Industry	Execution	Yes	Obsolete
BPMN	Industry	Graphical	Yes	Popular
BPQL	Industry	Diagnosis	Yes	Unfinished
BPRI	Industry	Diagnosis	Yes	Unfinished
ebXML BPSS	Industry	B2B	Yes	Popular
EDI	Industry	B2B	Yes	Stable
EPC	Academic	Graphical	No	Legacy
Petri nets	Academic	Theory/Graphical	N.A.	Popular
Π-Calculus	Academic	Theory/Execution	N.A.	Popular
Rosetta-Net	Industry	B2B	Yes	Popular
UBL	Industry	B2B	Yes	Stable
UML A.D.	Industry	Graphical	Yes	Popular
WSCI	Industry	Execution	Yes	Obsolete
WSCL	Industry	Execution	Yes	Obsolete
WS-CDL	Industry	Execution	Yes	Popular
WSFL	Industry	Execution	No	Obsolete
XLANG	Industry	Execution	No	Obsolete
XPDL	Industry	Execution	Yes	Stable
YAWL	Academic	Graphical/Execution	No	Stable

When all these aspects are provided, the system is called "BPMS". These aspects are provided if the system contains:

- an *orchestration engine*, that coordinates the sequencing of activities according to the designed flow and rules;
- a *business intelligence* and *analysis tools*, that analyze data produced during the executions. An example of this kind of tools is the Business Activity Monitoring (BAM) that provides real-time alerts for a proactive approach;
- a *rule engine*, that simplifies the changes to the process rules and provides more abstraction from the policies and from the decision tables, allowing more flexibility;
- a *repository* that stores process models, components, documents, business rules and all the information required for the correct execution of the process;
- tools for *simulation and optimization* of the process, that allow the designer to compare possible new process models with the current one in order to get an idea of the possible impact into the current production environment;
- an *integration* tool, that links the process model to other components in order to execute the process activities.

From a more pragmatic perspective, the infrastructure that seems to be the best candidate in achieving all the objectives indicated by BPM is the Service-Oriented Architecture (SOA) [54, 114, 117].

With the term SOA we refer to a model in which automation logic is decomposed into smaller, distinct units of logic. Collectively, these units constitute a larger piece of business logic; individually these can be distributed among different nodes. An example of such composition is presented in Fig. 2.8.

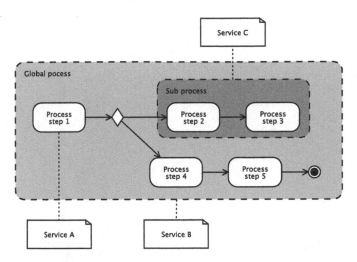

Fig. 2.8 Example of process handled by more than one service. Each service encapsulates a different amount of logic. This figure is inspired by Fig. 3.1 in [54].

In [140], a clear definition of "Business Service" is presented:

A discrete unit of business activity, with significance to the business, initiated in response to a business event, that can't be broken down into smaller units and still be meaningful (atomic, indivisible, or elementary).

This term indicates the so-called "internal requirements" of an Information System, in opposition to the "external" ones, identified as Use Cases: *a single case in which a specific actor will use a system to obtain a particular business service from one system*. In authors' opinion, this separation simplifies the identification of the requirements and can be considered a methodological approach to the identification of the components of the system.

In the context of SOA, one of the most promising technologies is represented by Web services. In this case, a Web service is going to represent a complex process that can span even more organizations. With the Web services composition, complex systems can be built according to the given process design; however, this is still a young discipline and industries should be more involved in the standardization process.

Fig 2.8. Example of process flow.

Chapter 3
Data Generated by Information Systems (and How to Get It)

The information systems described in Sect. 2.2, typically, record traces of each executed activity. All this information is then collected and stored in so called *log files*. There are several reasons for keeping a history of what happened: just to mention an example, it can be used to analyze errors that may occur (i.e., *debug*).

Process mining techniques use those logs as input for fact-based analyses, as reported in Chap. 5. Although most of the time information systems record data in a structured way, it may happen to deal with unstructured data. Information Extraction techniques are useful to give a structure to this kind of data. This chapter briefly reports the basic notions that characterize this problem.

3.1 Information Extraction from Unstructured Sources

The task of extracting information from unstructured sources is called Information Extraction (IE) [33, 42, 85]. These techniques can be considered as a type of Information Retrieval [103] in the sense that they aim at extracting automatically structured information from unstructured or semi-structured documents. Typically the core of IT techniques is based on the combination of Natural Language Processing tools, lexical resources and semantic constraints; so to extract, from text documents, important facts on some general entities (that the system needs to know *a priori*).

The information extraction systems can be divided into two main groups, based on their approach type: (*i*) learning systems; (*ii*) knowledge engineering systems. The first requires a certain number of already-annotated texts that are used as training set for some learning algorithm. The system can obtain a certain number of information that can then use with new texts. In the case of knowledge engineering systems, the responsibility for the accuracy of the extraction is assigned to the developer that has to construct a set of rules (starting from some example documents) and to develop the system.

A typical information extraction system encompasses the components presented in Fig. 3.1. The first component is responsible for the "tokenization": this phase consists

© Springer International Publishing Switzerland 2015
A. Burattin: *Process Mining Techniques in Business Environments*, LNBIP 207,
DOI 10.1007/978-3-319-17482-2_3

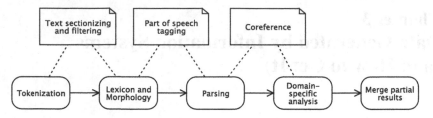

Fig. 3.1 Typical modules of an Information Extraction System, figure extracted from [85].

in splitting the text in sentences or, more generally, in "tokens". This problem cannot be solved for some type of languages (such as Chinese or Japanese). The second phase is composed of two parts: the morphological processing is fundamental for languages such German, where different nouns can be agglomerated into a single word. The lexical analysis consists in assigning to each token its lexical role in the sentence; the most important job is the identification of proper names. Typically, this phase extensively uses dictionary of words and roles. The parsing phase consists of removing from the text parts that one is not interested in and that are characterized by a particular structure (such as numbers, dates, ...). The coreference module is useful for identifying different ways of referring to the same entity. Typical problems handled by this module are:

1. *name-alias coreference*: names and possible variants;
2. *pronoun-antecedent coreference*: pronouns (like "he" or "she") are pointed to the correct entity;
3. *definite description coreference*: when a description is used instead of the name of an entity (e.g. "Linux" and "the Linus Torvalds' OS").

The domain-specific analysis is the most important step, but is also the most *ad hoc* one: it is necessary to define rules for each specific case in order to extract common behavior. These rules can be generated manually (Knowledge Engineering) or with machine learning approaches. The last step, the merging of partial results, consists in creating a single sentence from others, describing the same fact. This step is not necessary, but can help in presenting the outputs.

3.2 Evaluation with the F_1 Measure

The basic approaches used to evaluate the results of an information retrieval algorithm are based on the concepts of true/false positives/negatives. Table 3.1 presents these basic notions, by comparing the relevant and retrieved documents.

On top of these concepts, it is possible to define the concepts of *precision* and *recall*. The first represents the number of documents that are "retrieved" with respect to the number of documents that are actually relevant; the latter represents the number of relevant results that have been returned:

Table 3.1 Tabular representation of true/false positives/negatives. True negatives are colored in gray since they will not be considered.

	Relevant	Not relevant
Retrieved	True positives (tp)	False positives (fp)
Not retrieved	False negatives (fn)	True negatives (tn)

$$\text{Precision} = \frac{\# \text{ Relevant, Retrieved}}{\# \text{ Retrieved}} = \frac{\text{tp}}{\text{tp} + \text{fp}}$$

$$\text{Recall} = \frac{\# \text{ Relevant, Retrieved}}{\# \text{ Relevant}} = \frac{\text{tp}}{\text{tp} + \text{fn}}$$

One of the most commonly used metric for the evaluation of IR techniques is the F_1 [103]. This measure consists in the harmonic mean between precision and recall:

$$F_1 = 2 \cdot \frac{\text{Precision} \cdot \text{Recall}}{\text{Precision} + \text{Recall}}.$$

Chapter 4
Data Mining for Information System Data

Typically, the term "data mining" refers to the exploration and the analysis of large quantities of data, in order to discover meaningful patterns and rules.

In this book, we will make use of a particular kind of data mining technique known as *clustering*. We will present this technique up to the concepts and definitions needed later. Some other common data mining techniques will be presented as well but, for more information about these techniques and data mining in general, we refer to [126]. Typical data mining tasks, as reported in [13], are: *classification*, examining the features of a "new" object in order to assign it to one of the predefined set of classes (discrete outcomes); *estimation*, similar to classification, but deals with continuously valued outcomes (e.g. regression models or Neural Networks); *affinity grouping* (or *association rules*), determining how things can be grouped together (for example in a shopping cart at the supermarket); *clustering*, the task of segmenting a heterogeneous population into a certain number of homogeneous clusters (no predefined classes); *profiling*, simplification of the description of complicated databases in order to explain some behaviours (e.g. decision trees).

In the next sections one technique per task will be briefly presented.

4.1 Classification with Nearest Neighbor

The idea underpinning the nearest neighbor algorithm is that the properties of a certain instance are likely to be similar to the ones in its "neighborhood". To apply this idea, a distance function is necessary, in order to calculate the distance between any two objects. Typical distance functions are the "Manhattan distance" (d_M) and the "Euclidean distance" (d_E):

$$d_M(\mathbf{a}, \mathbf{b}) = \sum_{i=1}^{n} |\mathbf{b}_i - \mathbf{a}_i| \qquad d_E(\mathbf{a}, \mathbf{b}) = \sqrt{\sum_{i=1}^{n} (\mathbf{b}_i - \mathbf{a}_i)^2}$$

© Springer International Publishing Switzerland 2015
A. Burattin: *Process Mining Techniques in Business Environments*, LNBIP 207,
DOI 10.1007/978-3-319-17482-2_4

where **a** and **b** are two vectors in the n-dimensional space. Graphical representations of these two distances are reported in Fig. 4.1. This space is "populated" with all the pre-classified elements (examples): each object has a label that defines its class.

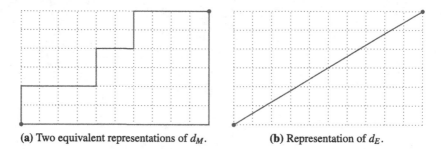

(a) Two equivalent representations of d_M.   (b) Representation of d_E.

Fig. 4.1 Graphical representations of "Manhattan distance" (d_M) and "Euclidean distance" (d_E) in a two dimensional space.

The idea is that the classification is obtained selecting the neighborhood of the new instance (typically, a parameter k indicates to select the first k nearest objects to the current one). Then the class of the new instance is selected as the most common class with respect to the current neighborhood. For example, if $k = 1$ then the class of the new instance is the same class of the nearest one already classified.

4.2 Neural Networks Applied to Estimation

Artificial Neural Networks are mathematical models that typically are represented as directed graphs where each vertex is a "neuron" that can be directly connected to other neurons. The function of a connection is to propagate the activation of one unit to the others. Each connection has a weight that determines the strength of the connection itself. There are three types of neurons: *input* (whose signals collect the external input), *output* (that will produce the result) and *hidden* (the ones that are between the input and the output neurons), as presented in the example of Fig. 4.2.

Each unit has an activation function that combines and transforms all its input values into signal for its output. A typical activation function is the one where the combination of all its input has to reach a certain threshold in order to increment the output. When the combination of input is above the threshold, the output is very high.

The training of the network aims at defining the weights of the connections between units (e.g. $w_{1,1}$ and $w_{n,m}$ in Fig. 4.2) so that, when a new instance is presented to the model, the output values can be calculated and returned as output. The main drawback of this approach is that the trained model is described only in terms of a set of weights that are not able to explain the training data.

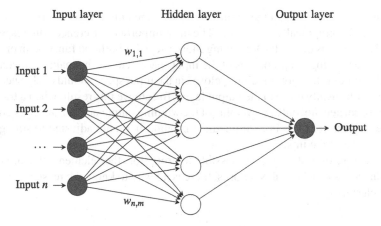

Input layer   Hidden layer   Output layer

Input 1 →

Input 2 →

··· →

Input n →

$w_{1,1}$

$w_{n,m}$

→ Output

Fig. 4.2 Example of Neural Network with an input, a hidden and an output layer. This network receives input from n neurons and produces output in one neuron.

4.3 Association Rules Extraction

An example of an association rule is *"if a customer purchases onions and potatoes than, the same customer, probably will purchase also a burger"*.

One of the most common algorithm to extract association rules is Apriori [4] that, given a set of transactions (called itemset), tries to find subsets of item that are common in many other itemsets (the basic idea is that a subset of a frequent itemset must also be a frequent itemset). These "frequent subsets" are incrementally extended until a termination condition is reached (i.e. there are no more possible extensions).

Starting from the frequent itemset B, the generation of the rules is done considering all the combination of subsets L and $H = B - L$. A rule $L \Rightarrow H$ is added if its confidence (i.e. how much H is observed, given the presence of L) is above a threshold.

4.4 Clustering

The term *clustering* refers to the problem of grouping together objects. In particular, elements of each group, called *cluster*, are supposed to be similar to each other, with respect to some distance measure (low intra-cluster distance). Moreover, elements belonging to different clusters are required to have low similarity (high inter-cluster distance).

4.4.1 Clustering with Self-organizing Maps

Self-organizing maps (SOM) can be considered as a variant of Neural Network. It has an input and an output layer, that consists of many units: each output unit is connected to all the units in the input layer. Since the output layer is organized as a "grid" it can be graphically visualized. The most important difference with respect to Neural Networks is that Self-Organizing Maps use neighborhood function in order to preserve the topological properties of the input space. This is the main characteristic of SOM: elements that are somehow "close" in the input space should enable neurons that are topologically close in the output layer. To achieve this result, when a training element is learned, not only the weights of the winning output neuron are adjusted, but also the weights for units in its immediate neighborhood are adjusted to strengthen their response to the input.

The training of a SOM makes possible to group together elements that are close but, as in the case of Neural Networks, there is no way to know the reasons that lead to such clustering.

4.4.2 Clustering with Hierarchical Clustering

Another technique to perform clustering is Hierarchical Clustering. In this case, the basic idea is to create an hierarchy of clusters. Clearly, a hierarchy of clusters is much more informative with respect to unrelated sets of clusters[1]. There are two possible approaches to implement Hierarchical Clustering: Hierarchical Agglomerative Clustering (HAC) and Hierarchical Divisive Clustering (HDC).

In HAC, at the beginning, each element constitutes a singleton cluster; and each iteration of the approach merges the closest clusters. The procedure ends when all the elements are agglomerated into a single cluster. HDC adopts the opposite approach: initially all elements belong to the same cluster so that each iteration splits the clusters until all elements constitute a singleton cluster.

The typical way to represent the result of Hierarchical Clustering is using a dendrogram, as shown in Fig. 4.3. Every time two elements are merged, the corresponding lines, on the dendrogram, are merged too. The numbers close to each merge represent the values of the *distance measure* for the two merged clusters.

To extract unrelated sets of clusters out of a dendrogram (as in flat clustering), it is necessary to set a *cut* (or *threshold*). This cut represents the maximum distance allowed to merge clusters. Therefore, only clusters with a distance lower than the threshold are considered as grouped. In the example of Fig. 4.3, two possible cuts are reported. Considering the cut at 0.5, these are the clusters extracted:

$$\{1\} \quad \{2, 3\} \quad \{4\} \quad \{5, 6\} \quad \{7\} \quad \{8\} \quad \{9\} \quad \{10\}.$$

Instead, the cut at 0.8 generates only two clusters:

$$\{1, 2, 3, 4, 5, 6\} \quad \{7, 8, 9, 10\}.$$

---

[1] In literature, sometimes, techniques generating a hierarchy of clusters and techniques generating unrelated sets of clusters are identified, respectively, as *hierarchical clustering* and *flat clustering*.

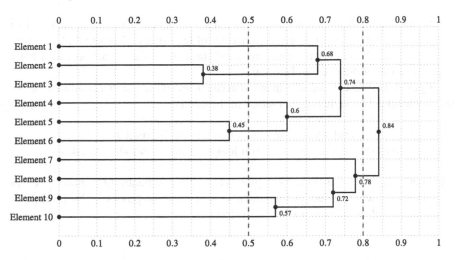

Fig. 4.3 Dendrogram example, with 10 elements. Two possible cuts are reported with red dotted lines (corresponding to values 0.5 and 0.8) (Color figure online).

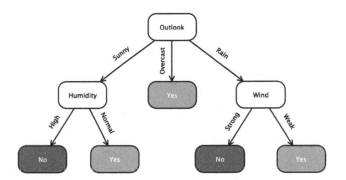

Fig. 4.4 A decision tree that can detect if the weather conditions allow to play tennis.

4.5 Profiling Using Decision Trees

Decision trees are data structure (trees, indeed) that are used to split collections of records in smaller subsets, by applying a sequence of simple decision rules.

The construction of a decision tree is performed top-down, choosing an attribute (the next best) and splitting the current set according to the values of the selected attribute. With each iterative division of the set, the elements of the resulting subsets become more and more similar one another. In order to select the "best" attribute a typical approach is to choose the one that splits the set into homogeneous subsets, however, there are different formulations of such definition.

An interesting characteristic of decision trees is that each path from the root to a leaf node can be considered as a conjunctions of tests on the attributes values; more paths towards the same leaf value encode disjunctions of conjunctions, so a logic representation of the tree can be obtained.

Figure 4.4 represents a decision tree that can detect if the weather conditions allow to play tennis. The logic representation of the tree is:

$$(\text{outlook} = \text{'sunny'} \wedge \text{humidity} = \text{'normal'}) \vee$$
$$(\text{outlook} = \text{'overcast'}) \vee (\text{outlook} = \text{'rain'} \wedge \text{wind} = \text{'weak'})$$

This peculiarity of decision trees improves their understandability by human beings and, at the same time, results fundamental in order to be processed by third-party systems (e.g. algorithms that need this information).

Chapter 5
Process Mining

Process mining is an emerging research field that originates from two areas: machine learning and data mining on one hand, process modelling on the other [148]. The general idea of process mining is to take, as input, some event data (e.g., event log files) and perform a fact-based analysis of process executions.

A typical example of possible fact-based analysis is *control-flow discovery*. The output entity of a control-flow discovery algorithm is a model of a "process", that is a description of how to perform an operation. More details on the definition of process have been presented in Sect. 2.1.

Typically, a process is described within the documentation of the company, in terms of "protocols" or "guidelines". However, these ways of representing the work (using natural language or ambiguous notations) are not required to be informative in terms of activities executed in the reality.

In order to discover how an industrial production process is actually performed, one could "follow" the product along the assembly line and see which steps are involved, their durations, bottlenecks, and so on. In a general context of business process, this observation is typically not possible due to a series of causes, for example:

- the process is not formalized (the knowledge about how to execute it, is tacitly spread among all workers involved);
- there are too many production lines, so that a single person is not able to completely follow the work;
- the process is not going to produce physical entries, but services of information;
- and other similar problems.

However, most of such processes are executed with the support of information systems, and these systems –typically– record all the operations that are performed in some "log files".

In Fig. 5.1, there is a representation of the main components involved in process mining and the interactions among them. First of all, the *incarnation* aspect (on the top right of the figure) represents the information system that supports the actual operational incarnation of the process. Such incarnation can be different from the

© Springer International Publishing Switzerland 2015   33
A. Burattin: *Process Mining Techniques in Business Environments*, LNBIP 207,
DOI 10.1007/978-3-319-17482-2_5

ideal process definition and describes the actual process, as it is being executed. As we said, the information system records all the operations that are executed in some event logs. These *observations* are a fundamental requirement for the analysis using *process mining* techniques. Such techniques can be considered as the way of relating event logs (*what is happening*) to the analytical model of the process (*what is supposed to happen*). Analytical models (depicted on the left side of Fig. 5.1, into the imagination aspect) are supposed to describe the process, but an operational model is needed to add the detailed and concrete information that are necessary for its execution.

"Process mining" refers to the task of discovering, monitoring and improving real processes (as they are observed in the event logs) with the extraction of knowledge from the log files. It is possible to distinguish at least three types of mining (presented in Fig. 5.1 as grey boxes with arrows describing the interaction with other components):

1. **Control-flow discovery** aims at construction of a model of the process, starting from the logs (an *a priori* model is not required) [168];
2. **Conformance** analysis: starting from an *a priori* model, conformance algorithms try to fit the observations of the actual performed process in the original model and *vice versa*, as, for example, presented in [132];
3. **Extension** of a model, already available, in order to add information on the decision points (as presented in [131]) or on the performance of the activities.

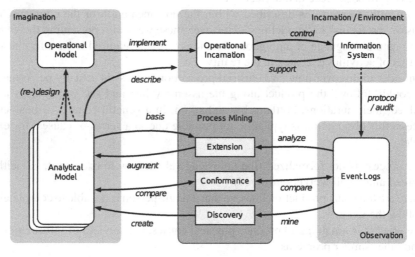

Fig. 5.1 Representation of the three main perspectives of process mining, as presented in Fig. 4.1 of [72].

Closely to the possible ways of performing the mining there are the three possible perspectives: the *control-flow* (that represents the ordering of the activities); the *organizational* or *social* (that focuses on which performers are involved and how are they related) and the *case* (how data elements, related to the process instance, evolve).

5.1 Process Mining as Control-Flow Discovery

This section provides some information on the state of the art for what concerns process mining and, in particular, control-flow discovery algorithms. Since the idea of this section is to provide a "history" of the field, the contributions are presented according to chronological order.

Figure 5.2 depicts a timeline, where each point indicates one or more approaches published. It is worthwhile to notice the "evolution" of the algorithms (the firsts generate simple models, without considering noise as a problem; the latest ones produce complex models and try to deal many problems). This is not intended as an exhaustive list of all control-flow algorithms, so it contains only the most important ones.

Fig. 5.2 Timeline with the control-flow discovery algorithms discussed in the state of the art of this work.

Cook and Wolf

The first work in the field of process mining is recognized in the Ph.D. Thesis of Jonathan Cook [35] and in other works co-authored with Alexander Wolf et al. [36–40]. The main contribution of the work consists of three different algorithms based on three different approaches: RNet, KTail and Markov.

All those algorithms are implemented in a tool called Balboa [39]. Balboa is a tool for analyzing data generated from software processes: the basic idea is to work on "events databases".

They define an "event" as an action that can be identified and that is instantaneous (e.g. the invocation, by a user, of a software). For this reason, an activity that lasts for a certain period of time is described in terms of two events ("start" and "end").

RNet. The RNet algorithm is based on a Recurrent Neural Network [109]. This type of networks can be seen as a graph with cycles where each node is a "network unit" (such that, given an input, it can calculate the corresponding output) and each node is weighted (initially, the weight is low). This network will calculate the output values for all its components at time t and then this output is used as input for the network at time $t + 1$. With such topology it is possible to model the behavior of an automaton: the final result is not computed on the basis of the input only, but also on the basis of the previous activity of the hidden neurons. The foremost advantage of such technique is that it is entirely statistical, so it is very robust to noise. The main drawback, however, is that, in order to be entirely exploited, this approach requires also "negative examples" (examples that can not be generated by the process); unfortunately, in real cases, it is very hard to obtain this information.

KTail. The second approach, KTail, differently from the previous one, is entirely algorithmic. The basic notion, in the whole system, is that a "state" is defined on the basis of all the possible future behavior. For example, two strings (as series of activities) can have a common prefix and, at a certain character, they can diverge one from the other. In this case, we have two strings with "a common history but different futures". Conversely, if two different stories shares the same future then they belong to the same equivalence class, that represents the automaton states (actually, the final model constructed is an automaton).

Markov. The last approach presented into the Balboa framework is called Markov. This is a hybrid approach, both statistical and algorithmic. A Markov model is used to find the probabilities of all the possible productions and, algorithmically, those probabilities are converted into an automaton. The assumptions made by this approach are the following:

- the number of states of the current process is finite;
- at each time, the probability that the process is in one of its states depends on the current state (Markov property);
- the transition probabilities do not change over time;
- the starting state of the process is probabilistically determined.

The last approach above-mentioned, Markov, seems to be the one with the best results, also because of the probabilistic approach, that allows the procedure to be noise tolerant.

Agrawal et al.

The approach developed by R. Agrawal, D. Gunopulos and F. Laymann [3] is considered the first process mining algorithm in the context of BPM. In particular, the aim of their procedure is to generate a directed graph $G = (V, E)$, where V is the set of process activities and E represents the dependencies among them. Initially, $E = \emptyset$, and the algorithm will try to build a series of boolean function fs such that:

$$\forall\ (u, v) \in E \quad f_{(u,v)} : \mathbb{N}^k \to \{0, 1\}$$

that, starting from the output of the activity u, indicates if v can be the next one.

In this approach, a process execution log is a set of tuples (P, A, E, T, O) where P is the name of the process; A is the name of the activity; $E \in \{start, end\}$ is the event type; T is the time the action took place; $O = o(A)$ is the output produced by the activity A, if $E = end$, otherwise it is a null value.

Dependencies between activities are defined in a straightforward manner: B depends on A if, observing the real executions, it is very common that A is followed by B and never the *vice versa*. Since all the activities are thought as a time interval, it is important to point the meaning of "A is followed by B". The definition describes two possible cases: *(i)* B starts after A is finished; *(ii)* an activity C exists such that C follows A and B follows C.

The whole procedure can be divided into two main phases. The first part deals with the identification of the dependencies between activities. This is done observing the logs and adding the edge (u, v) to E every time u ends before v starts. A basic support for the noise is provided by counting the number of times every edge is observed and excluding from the final model all the edges that do not reach a threshold (parameter of the procedure). A problem that can emerge is the configuration of the threshold value. A solution, proposed in the paper, is to convert the threshold into an "error probability" $\varepsilon < \frac{1}{2}$. With this probability, it is possible to calculate the minimum number of observations required.

The second step of the approach concerns the definition of the conditions required in order to make edges followed each other. The procedure presented in the paper defines the function $f_{(u,v)}$ which uses the output values produced as output by the activities: for all the executions of activity u if, in the same process instance, activity v is executed, then the value $o(u)$ is used as a "positive example". The set of values produced can be used as *training set* for a classification task. The paper suggests to use decision trees [109] for learning simple rules that can be also understood by human beings.

Herbst and Karagiannis

In the approach presented by Herbst and Karagiannis in [77-79] a workflow W is defined as $W = (V_W, t_W, f_W, R_W, g_W, P_W)$ where:

- $V_W = \{v_1, \ldots, v_n\}$ is a set of nodes;
- $t_W(v_i) \in \{Start, Activity, Decision, Split, Join, End\}$ indicates the "type" of a node;
- $f_W : V_{ACT} \to A$ is a function for the assignment of nodes to activities;
- $R_W \subseteq (V_W \times V_W)$ is a set of edges, where each edge represents a successor relation;
- $g_W : R_W \to COND$ are transition conditions;
- $P_W : R_W \to [0, 1]$ are transition probabilities.

with V_X, for $X \in \{Start, Activity, Decision, Split, Join, End\}$, that denotes the subset $V_X \subseteq V_W$ of all nodes of type X.

With these definitions, the mining algorithm aims at discovering a "good approximation" of the workflow that originated the observations.

The workflow is expressed in terms of Stochastic Activity Graph (SAG), that is a directed graph where each node is associated to an activity (more nodes referring to the same activity are allowed) and each edge (that represents a possible way of continuing the process) is "decorated" with its probability. Additionally, each node must be reachable from the start and the end must be reachable from every node. The procedure can be divided in two phases: the identification of a SAG, that is "consistent" with the set of process instances observed (the log), and the transformation of the SAG into a workflow model.

In order to identify the SAG, the procedure described is very similar to the one of Agrawal et at. [3], and it is based on the creation of a node for each activity observed; the generation of edges is performed according to the dependencies observed in the log.

Hwang and Yang

In the solution proposed by S. Y. Hwang e W. S. Yang [82], each activity is described as a time interval. In particular, an activity is composed of three possible sub-events:

1. a *start event* that determines the beginning of the activity;
2. an *end event* that identifies the activity conclusion;
3. the possible *write event* used for the identification of the writings of the output produced by the activity.

All these possible events are atomic, so it is not possible to observe two of them at the same time.

The *start* and the *end event* are recorded in a triple that contains the name of the activity, the case identifier and the execution time. The *write events* present the same fields of the other two but, in addition, they contain also information on the written variables and their values.

Two activities, belonging to the same instance, can be described as "disjoint" or "overlapped". They are disjoint if one starts after the end of the other; they are overlapped if they are not disjoint. The aim of the approach is to find all the couples of disjoint activities (X, Y) such that X is directly followed by Y: all those couples are the candidates for the generation of the final dependency graph that represents the discovered process. Another constructed set is the one with all the overlapped activities. Starting from the assumption that two activities overlapped are not in a dependency relation, the final model is constructed adding an edge between two activities that are observed in direct succession and not overlapped.

In order to identify the noise in the log, the proposed approach describes other relations among activities and it considers only the observations that exceed the value given as threshold. Moreover the output of each activity is proposed to be used for the definition of the conditions for the splits.

Schimm

In the work by Guido Schimm [135, 136, 168], there is a definition of *trace* as a set of events, according to the described life cycle of the activities. Such life cycle, can be considered quite general and is proposed in Fig. 5.3. In the work, however, only the events *Start* and *Complete* are considered, but these are sufficient for the identification of parallelisms.

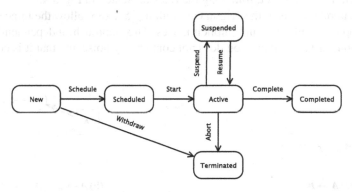

Fig. 5.3 Finite state machine for the life cycle of an activity, as presented in Fig. 1 of [136].

The language used for the representation of the resulting process is a block based language [92] where every block can be: a sequence, a parallel operator, or an alternative operator. An example of a "mined model", with activities a, b, c and d is:

$$\mathscr{A}(\mathscr{P}(\mathscr{S}(a, b), (c, d)), \mathscr{S}(\mathscr{P}(c, \mathscr{S}(b, a)), d))$$

where $\mathscr{S}$ identifies the sequence operator, $\mathscr{P}$ the parallel and $\mathscr{A}$ the alternative. Of course, the same process can be also graphically represented.

The procedure starts finding all the precedence relations, also the pseudo ones (dependencies that maybe do not exist in the original model, that are due to some random behavior such as delays); and then ends converting all them into the given model. It is interesting to note that this approach aims only at describing the behavior contained in the model, without any generalization.

There is a tool that implements the above-mentioned approach, and that can be downloaded for free from the Internet[1].

---

[1] At the website: http://www.processmining.de.

Van der Aalst et al.

The work by van der Aalst et al. [162, 169] is focused on the generation of a Petri net model that can describe the log. The idea, formalized in an algorithm called "Alpha", is that some relations – if observed in the log – can be combined together in order to construct the final model. These relations, between activities a and b, are:

- the direct succession $a > b$, when, in the log, sometime a compares before b;
- the causal dependency (or follow) $a \rightarrow b$, when $a > b$ and $b \ngtr a$;
- the parallelism $a \| b$, when $a > b$ and $b > a$;
- uncorrelation #, when $a \ngtr b$ and $b \ngtr a$.

Given sets containing all the relations observed into the log, it is possible to combine them generating a Petri net, following the rules presented in Fig. 5.4.

Some improvements to the original algorithm [151, 183] allow the mine of short loops (loop of length one) and implicit places. This approach, independently from its extensions, assumes that a log does not contain any noise and that it is complete

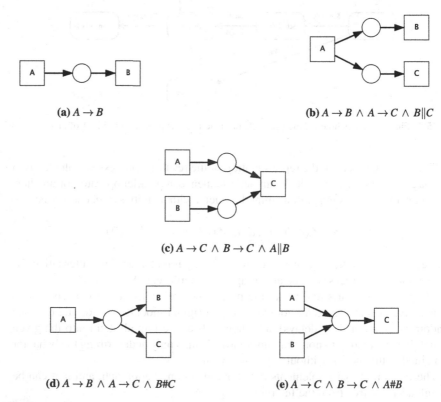

(a) $A \rightarrow B$

(b) $A \rightarrow B \wedge A \rightarrow C \wedge B \| C$

(c) $A \rightarrow C \wedge B \rightarrow C \wedge A \| B$

(d) $A \rightarrow B \wedge A \rightarrow C \wedge B\#C$

(e) $A \rightarrow C \wedge B \rightarrow C \wedge A\#B$

Fig. 5.4 Basic ideas for the translation of set of relations into Petri net components. Each component caption contains the logic proposition that must be satisfied.

with respect to the follow relation so, if an activity A is expected to be in the final model directly before B, it is necessary that the relation $A \rightarrow B$ is observed, in the log, at least one time. Moreover, as can be deduced, there is no statistical analysis of the frequency of the activities. These three observations prevent the current approach to be applied in any real context.

Golani and Pinter

As in other approaches, the one presented by Mati Golani and Shlomit Pinter [66, 123] considers each activity as a not instantaneous event, in the sense that it is composed of a "*start*" and "*end*" event. In order to reconstruct the model, given two activities a_i and a_j contained into a log, the procedure defines the dependency of a_i on a_j iff, whenever a_i appears in some execution in the log, a_j must appear in that execution sometime earlier and the termination time of a_j is smaller than the starting time of a_i. The notion of time interval is crucial for the application of the procedure since it analyses the overlaps of time intervals. The presented approach is quite close to the one by Agrawal et al.

Weijters et al.

The control-flow discovery approach by Weijters et al. [165, 167] is called "Heuristics Miner". This algorithm can be considered as an extension of the one by van der Aalst et al. (2002): in both cases the idea is to identify sets of relations from the log and then build the final model on the basis of such observed relations. The main difference between the approaches is the usage of statistical measures (together with acceptance thresholds, parameters of the algorithm) for the determination of the presence of such relations.

The algorithm can be divided into three main phases: the identification of the graph of the dependencies among activities; the identification of the type of the split/join (each of them can be an AND or a XOR split); the identification of the "long distance dependencies". An example of measure calculated by the algorithm is the "dependency measure" that calculates the likelihood of a dependency between an activity a and b:

$$a \Rightarrow b = \frac{|a > b| - |b > a|}{|a > b| + |b > a| + 1}$$

where $|a > b|$ indicates the number of times that the activity a is directly followed by b into the log. The possible values of $a \Rightarrow b$ are in the range -1 and 1 (excluded); its absolute value indicates the probability of the dependency and its sign indicates the "direction" of the dependency (a depends on b or the other way around). Another measure is the "AND-measure" which is used to discriminate between AND and XOR splits:

$$a \Rightarrow (b \wedge c) = \frac{|b > c| + |c > b|}{|a > b| + |a > c| + 1}$$

If two dependencies are observed, e.g. $a \rightarrow b$ and $a \rightarrow c$, it is necessary to discriminate if a is and AND or a XOR split. The above-written formula is used for this purpose: if the resulting value is above a given threshold (parameter of the algorithm), then a is considered as an AND split, otherwise as XOR.

This is one of the most used approaches in real-case applications, since it is able to deal with noise and is able to produce generalized relations.

Greco et al.

The process mining approach by Gianluigi Greco et al. [68, 69] aims at creating a hierarchy of process models with different levels of abstraction.

The approach is divided into two steps: in the first one, the traces are clustered with an iterative partitioning of the log. In order to perform the clustering, some features are extracted from the log using a procedure that identifies the "frequent itemset" of sequence of activities among all the traces. Once the clustering is completed, a hierarchy (i.e. a tree) is built containing all the clusters: each node is supposed to be an abstraction of its children, so that different processes are abstracted into a common parent.

Van Dongen et al.

The approach by van Dongen et al. [175, 176] aims at generating Event-driven Process Chains (EPC) [50]; this notation does not require a strong formal framework because, among other things, the notation does not rigidly distinguish between output flows and control-flows or between places and transitions, as these often appear in a consolidated manner. An example of EPC is proposed in Fig. 5.5.

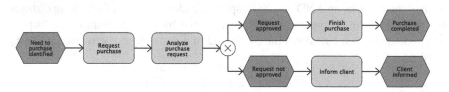

Fig. 5.5 An example of EPC. In this case, diamonds identify events; rounded rectangles represent functions and crossed circles identify connectors.

In this approach, a model (actually, it is just a partial order, with no AND or XOR choices) is generated for each log trace. After the set of "models" is completely generated (so that all the traces have been observed), these are aggregated in a single model. The aggregation is constructed according to some rules but, intuitively, can be considered as the sum of the behaviors observed in all the "trace model".

Alves de Medeiros et al.

The approach presented by Ana Karla Alves de Medeiros and Wil van der Aalst [44, 153] uses genetic algorithms [8, 109].

In the first step, a random "population" of initial processes is created. Each individual of this population is analyzed in order to check how much it "fits" the given log (in this approach, the fitness criterion is fundamental). After that, the population evolves according to genetic operators: *crossover* helps combining two individuals; *mutation* modifies a single model.

With this approach, the algorithm iteratively improves the population, until a suitable candidate is found ("stop criterion"). This approach is extremely powerful and can extract a huge number of models, but the main drawback is its huge complexity.

Günter et al.

In [72, 74] Christian Günter and Wil van der Aalst present their new idea for handling "spaghetti-processes". These processes are characterized by an incredible level of flexibility that reduces their "structureness". Figure 5.6 presents a mined model (and an enlarged portion of it) based on the executions of unstructured process: there are many possible paths, thus an approach that tries to describe the complete process is not correct.

Usually, they propose a metaphor with road maps: in a map that presents the entire country it does not make any sense to present all the streets of all the cities; instead, a city-map should propose all those small routes. The same idea of "abstraction" is applied to process mining, and is based on two concepts: *significance* (behaviors important in order to describe the log) and *correlation* (two behaviors close one to the other). With these two concepts, it is possible to produce the final model considering these heuristics:

- highly significant behaviors need to be preserved;
- less significant, but highly correlated behavior should be aggregated;
- less significant and lowly correlated behavior should be removed from the model.

The result of their mining approach is a procedure that creates an "interactive" dependency graph. According to the user's requirements, it is possible to "zoom in", adding more details, or "zoom out", creating clusters of activities and removing edges.

Goedertier et al.

In [64], Goedertier et al. present the problem of process discovery from a new point of view: using not only the "classical" observations of executed activities, but also with sequence of activities that are not possible.

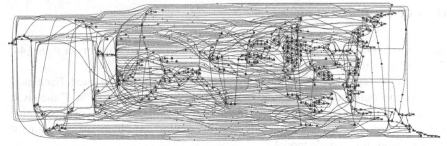

(a) The entire mined process model.

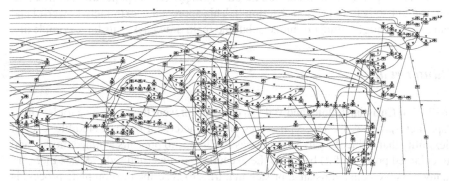

(b) A fraction of the process.

Fig. 5.6 Example of a mined "spaghetti model", extracted from [74].

Essentially, all the real-world business logs contain only "positive events": behavior that the system does not allow, typically, are not recorded. This aspect limits the process discovery to a setting of unsupervised learning. For this reason, authors decided to artificially generate negative events (behavior not allowed by the model) following these steps:

1. process instances with the same sequence of activities are grouped together (to improve efficiency);
2. completeness assumption: behavior that does not occur in the log should not be learned;
3. induce negative events checking all the possible parallel variants of the given sequence (permutation of some activities).

Once the log (with positive and negative events) is generated, the system learns the precondition of each activity (as a classification problem, using TILDE: given an activity in a particular time, detect if a positive or negative event takes place). The set of precondition is then transformed into a Petri net on the basis of correspondences between language constructs and Petri net patterns (these rules are similar to the Alpha miner rules presented in Fig. 5.4).

Maggi et al.

The approach by Maggi et al. [101] can be used to mine a declarative process model (expresses using Declare language, see Subsect. 2.1.4).

The basic idea of the approach is to ask the user which kind of constraints to mine (i.e., the *Declare templates*). Once the user has inserted all the templates, the system builds the complete list with the actual constraints, by applying each template to all the possible combinations of activities. All constraints are checked against the log: if the log violates one constraint (i.e. it does not hold for at least one trace), it is removed and not considered anymore. Once the procedure has completed the analysis of all the constraints, a Declare model can be built (as the union of all the holding constraints).

The described procedure provides two parameters (*Percentage of Events* and *Percentage of Instances*) that are useful to define, respectively, the activities to be used to generate the candidate constraints, and to specify the number of traces that a constraint is allowed to violate to be still considered in the final model. These two parameters are useful to deal with noise in the data.

Other Approaches

A set of other process mining approaches is based on the Theory of Regions (in Petri nets) [11, 171, 172]. Their idea is to construct a finite state automaton with all the possible states observable into the log and then transform it into a Petri net using the Theory of Regions (where "region" of states with the same input/output are collapsed into a single transition).

Recent approaches are now focusing on the discovery of *process trees* (i.e., a block-structured language) [19, 20, 95]: they are able to discover models with specific guarantees such as soundness or perfect fitness.

5.2 Other Perspectives of Process Mining

It is wrong to consider process mining as just control-flow discovery [83]: instead, there are other perspectives that it is useful to consider. All the approaches described in the following subsections can provide interesting insights on the process under analysis, even if they do not consider the control-flow discovery as the main problem.

5.2.1 Organizational Perspective

An important side of process mining is the social mining [157, 158] (i.e. the organizational perspective of process mining). The social perspective of process mining consists in performing Social Network Analysis (SNA) on data generated from business processes. In particular, it is possible to distinguish between two types of approaches:

sociocentric and *egocentric* approaches. The first consists in analyzing the network of connections as a whole, considering the complete set of interactions within a group of persons. The second approach concentrates on a particular individual and analyzes her or his set of interactions with other persons.

These relationships are established according to four metrics: *(i)* causality; *(ii)* metrics on joint cases (instance of processes where two individuals operates); *(iii)* metrics on joint activities (activities performed by the same persons); *(iv)* metrics based on special event types (e.g. somebody suspends an activity and another resumes it).

5.2.2 Conformance Checking

Another important aspect of process mining is conformance checking. This problem is completely different from both control-flow discovery and SNA: the "input" of conformance checking consists in a log and a process model (it can be defined "by hand" or it can be discovered), and it aims at compare the observed behavior (i.e. the log) with what is expected (i.e. the model) in order to discover discrepancies.

It is possible to instantiate the conformance checking in two activities: *(i) business alignment* [128, 147] and *(ii) auditing* [62]. The aim of the first is to verify that the process model and the log are "well aligned". The second one tries to evaluate the current executions of the process with respect to "boundaries" (given by managers, laws, etc.).

Conformance checking approaches start becoming available for declarative process models too, as described in [96].

5.2.3 Data Perspective

When considering a process model, it can be interesting to consider also the "data perspective". This term refers to the integration of the control-flow perspective with other "ornamental" data.

For example, in [130, 131], authors describe a procedure that is able to decorate the branches of a XOR-split (for example, of a Petri net) with the corresponding "business conditions" that are required in order to follow that path. The procedure uses data recorded into the log that are related to particular activities and, using decision trees (Sect. 4.5), it extracts a logic proposition that holds on each branch of the XOR-split.

5.3 Performance Evaluation of Process Mining Algorithm

Every time a new process mining algorithm is proposed, an important question emerges: how is it possible to compare the new algorithm against the others, already available in the literature? Is the new approach "better" with respect to the others? Which are the performances of the new algorithm?

The main point is that, typically, the task of mining a log is an "offline activity", so the optimization of the resources required (in terms of memory and CPU power) is not the main goal. For these mining algorithms, it is more important to achieve "precise" and "reliable" results.

The main reason behind the creation of a new process mining algorithm is that the current ones do not produce the expected results or, in other terms, that the data analyzed contain information that are different from the required ones. In order to compare the performances of the two control-flow algorithms, the typical approach lies in comparing the original process model (the one that, executed, generates the given log) with the mined one. A graphical representation of such "evaluation process" is presented in Fig. 5.7.

Fig. 5.7 Typical "evaluation process" adopted for process mining (control-flow discovery) algorithms.

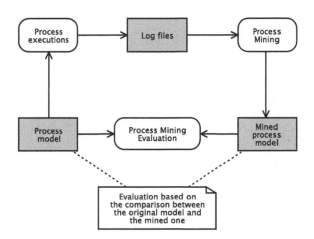

In the field of data mining, it is very common to compare new algorithms against some published datasets so all the other researchers can obtain the results claimed by the algorithm creator. Unfortunately, nothing similar exists for process mining: the real "owner" of business processes (and thus of logs) are companies that, typically, are reluctant to publicly distribute their own business process data: in this way, it is difficult to build up a suite of publicly available business process logs for evaluation purposes. Of course, the lack of extended process mining benchmarks is a serious obstacle for the development of new and more effective process mining algorithms. A way around this problem is to try to generate "realistic" business process models together with their execution logs. A first attempt to do such a models and logs generator is presented in [25, 26].

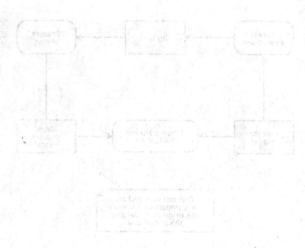

Chapter 6
Quality Criteria in Process Mining

When it is necessary to evaluate the result of a control-flow discovery algorithm, a good idea is to split the evaluation into different aspects. In [129], four dimensions are presented:

1. the **fitness** indicates how much of the observed behavior is captured by the process model;
2. the **precision** points out if a process is overly general (a model that can generate many more sequences of activities with respect to the observations in the log);
3. the **generalization** denotes if a model is overly precise (a model that can produce only the sequence of activities observed in the log, with no variation allowed);
4. the **structure** indicates the difficulties in understanding the process (of course, this measure depends on the language used for representing the model and defines the difficulties in reading it).

Table 6.1 Example of log traces, generated from the executions of the process presented in Fig. 6.1(a).

Log trace	Frequency
ABDEI	1207
ACDGHFI	145
ACGDHFI	56
ACHDFI	28
ACDHFI	23

These dimensions can be used for the identification of the aspects highlighted in a model. For example, in Fig. 6.1 four processes are displayed with different levels for the different evaluation dimensions. Suppose, as reference model, the one in Fig. 6.1(a), and assume that a log that it can generate is presented in Table 6.1. The process in Fig. 6.1(b) is called "flower model" and allows any possible sequence of activities: so, essentially, it does not define an order among them. For this reason, even if it has high fitness, generalization and structure, it has very low precision. The process Fig. 6.1(c) is just the most frequent sequence observed in the log, so it has

© Springer International Publishing Switzerland 2015
A. Burattin: *Process Mining Techniques in Business Environments*, LNBIP 207,
DOI 10.1007/978-3-319-17482-2_6

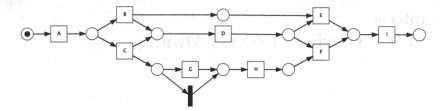

(a) The reference model, good balance among all dimensions

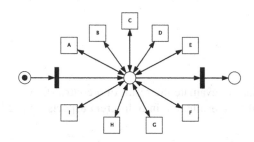

(b) Model with low precision but high fitness, generalization and structure

(c) Model with low fitness and generalization but high precision and structure

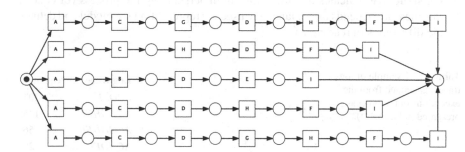

(d) Model with low generalization and structure but high fitness and precision

Fig. 6.1 Four process where different dimensions are pointed out (inspired by Fig. 2 of [129]). The **(a)** model represents the original process, that generates the log of Table 6.1; in this case all the dimensions are correctly highlighted; **(b)** is a model with a low fitness; **(c)** has low precision and **(d)** has low generalization and structure.

low fitness and generalization, but medium precision and high structure. The process Fig. 6.1(d) is a "complete" model, where all the possible behaviours observed in the log can be reproduce without any flexibility. This model has low generalization and structure but high fitness and precision.

In the remaining part of this chapter some metrics are presented. In particular, it is possible to distinguish between model-to-log metrics, which compare a model with a log, and model-to-model metrics which compare two models.

6.1 Model-to-Log Metrics

These metrics aim at comparing log with the process model that — using process mining techniques — has been derived.

Completeness (to quantify fitness) [70] defines which percentage of the **traces** in a log can also be generated by the model.

Parsing Measure (to quantify fitness) [167] is defined as the number of correct parsed traces divided by the number of traces in the event log.

Continuous Parsing Measure (to quantify fitness) [167] is a measure that is based on the number of successfully parsed events instead of the number of parsed traces.

Fitness (to quantify fitness) [132] considers also the "problems" happened during replay (e.g. missing or remaining tokens in a Petri net), so that actions that can't be activated are punished as the action that remains active in an improper way.

Completeness (PF Complete) (to quantify fitness) [44] very close to the *Fitness*, takes into account trace frequency as weights when the log is replayed.

Soundness (to quantify precision/generalization) [70] calculates the percentage of traces that can be generated by a model and that are in a log (so, the log should contain all the possible traces).

Behavioral Appropriateness (to quantify precision/generalization) [132] evaluates how much behavior is allowed by the model but never used in the log of observed executions.

ETC Precision (to quantify precision/generalization) [110] evaluates the precision by counting the number of times that the model deviates from the log (by considering the possible "escaping edges").

Structural Appropriateness (to quantify structure) [132] measures if a model is less compact than the necessary, so extra alternative duplicated tasks (or redundant and hidden tasks) are punished.

6.2 Model-to-Model Metrics

The following metrics aim at comparing two models, one against the other.

Label Matching Similarity [49] is based on a pairwise comparison of node labels: an optimal mapping equivalence between the nodes is calculated and the score is the sum of all label similarity of matched pairs of nodes.

Structural Similarity [49] measures the "graph edit distance", that is the minimum number of graph edit operations (e.g. node deletion or insertion, node substitution, and edge deletion or insertion) that are necessary to get from one graph to the other.

Dependency Difference Metric [7] counts the number of edge discrepancies between two dependency graph (binary tuple of nodes and edges).

Similarity measure for restricted workflows (graph edit distance) [108] another edit distance measure, based on dependency graph of the model.

Process Similarity (High-level Change Operations) [97] this measure counts the changes required to transform a process into another oner one, with "high level" changes (not adding or removing edges, but "adding activity between two", and so on).

Behavioral Similarity (Cosine Similarity for Causal Footprints) [106] is based on the distance between causal footprints (graph describing the possible behaviors of a process) and, in particular, calculates the cosine of the angle between the two footprints vectors (representations of the causal footprint graph).

Behavioral Profile Metric [93] compares two processes in terms of the corresponding behavioral profiles (characteristics of a process expressed in terms of relations between activity pairs).

Chapter 7
Event Streams

Process mining approaches have always been defined for *static* (i.e., finite, and not evolving) log files. However, nowadays, with the advent of *smart devices*, it is more relevant to begin dealing with so called *event streams*.

Part IV of the book will entirely be dedicated to process mining applied to event streams. This chapter provides the basic notions and the state of the art of such discipline.

7.1 Data Streams

A data stream is defined as a *"real-time, continuous, ordered sequence of items"* [65]. The ordering of the data items is expressed implicitly by the arrival timestamp of each item. Algorithms that are supposed to interact with data streams must respect some requirements, such as:

1. it is impossible to store the complete stream;
2. backtracking over a data stream is not feasible, so algorithms are required to make only one pass over data;
3. it is important to quickly adapt the model to cope with unusual data values;
4. the approach must deal with variable system conditions, such as fluctuating stream rates.

Due to these requirements, algorithms for data streams mining are divided into two categories: data and task based [61]. The idea of the first ones is to use only a fragment of the entire dataset (by reducing the data into a smaller representation). The idea of the latter approach is to modify existing techniques (or invent new ones) to achieve solutions efficient in terms of time and space.

© Springer International Publishing Switzerland 2015
A. Burattin: *Process Mining Techniques in Business Environments*, LNBIP 207,
DOI 10.1007/978-3-319-17482-2_7

7.1.1 Data-Based Mining

The main "data based" techniques are: *sampling*, *load shedding*, *sketching* and *aggregation*. All these are based on the idea of randomly select items or stream portions. The main drawback is that, since the dataset size is unknown, it is hard to define the number of items to collect; moreover it is possible that some of the items that were ignored were actually interesting and meaningful. Other approaches, like aggregation, are slightly different: since they are based on summarization techniques, in this case, the idea is to consider measures such as mean and variance; with these approaches, problems arise when the data distribution contains many fluctuations.

7.1.2 Task-Based Mining

The main "task based" techniques are: approximation algorithms, sliding window and algorithm output granularity. Approximation algorithms aim to extract an approximate solution. It is possible to define error bounds on the procedure. This way, one obtains an "accuracy measure". The basic idea of sliding window is that users are more interested in most recent data, thus the analysis is performed giving more importance to recent data, and considering only summarization of the old ones. The main characteristic of "algorithm output granularity" is the ability to adapt the analysis to resource availability.

7.2 Common Stream Mining Approaches

The task of mining data stream is typically focused on specific types of algorithms [2, 61, 184]. In particular, there are techniques for: clustering; classification; frequency counting; time series analysis and change diagnosis (concept drift detection). All these techniques cope with very specific problems and can hardly be adapted to any process mining problem. However, as this book shows, it is possible to reuse some stream mining principles into process mining sub-problems, which can be solved with the available algorithms.

7.3 Stream Mining and Process Mining

Over the last decade dozens of process mining and, in particular, control-flow discovery techniques have been proposed, e.g., the Heuristics Miner [165]. However, all these techniques work on a full event log and not on streaming data. Few works in process mining literature touch issues related to mining event data streams.

In [88, 89], the authors focus on incremental workflow mining and *task mining* (i.e. the identification of the activities starting from the documents accessed by users). The basic intuition is to mine process instances as soon as they are observed; each new model is then merged with the previous one so to refine the global process representation. The approach described is thought to deal with the incremental process refinement, based on logs generated from version management systems. However, as authors state, only the initial idea is sketched.

An approach for mining legacy systems is described in [86]. In particular, after the introduction of monitoring statements into the legacy code, an incremental process mining approach is presented. The idea is to apply the same heuristics of the Heuristics Miner into the process instances and to add these data into an AVL tree (this kind of trees are used to find the best holding relations). Actually, this technique operates on "log fragments", and not on single events, so it is not really suitable for an online setting. Moreover, heuristics are based on frequencies, so they must be computed with respect to a set of traces and, again, this is not suitable for the settings with streaming event data.

An interesting contribution to the analysis of evolving processes is given in the paper by Bose et al. [17]. The proposed approach, based on statistical hypothesis tests, aims at detecting *concept drift*, i.e. the changes in event logs, and at identifying the regions of change in a process.

Solé and Carmona, in [141], describe an incremental approach for translating transition systems into Petri nets. This translation is performed using Region Theory. The approach solves the problem of complexity of the translation, by splitting the log into several parts; applying the Region Theory to each of them and then combining all them. These regions are finally converted into Petri net.

Part II
Obstacles to Process Mining in Practice

This part introduces the problems that emerge when process mining is applied in real-world environments.

The prime example of such obstacles are infrequent or "noisy" behaviors, which occur in real life processes as a deviation from normal process flow. It is desirable to automatically discover models that explicitly exclude such infrequent behaviors, to ease the understanding of the process at hand, however not all the algorithms can effectively deal with such problem.

In general, however, it is possible to observe problems before, during and after the actual mining phase. Each of these periods will be commented in this part.

Chapter 8
Obstacles to Applying Process Mining in Practice

Process mining and, in particular, control-flow discovery, have made large advancements in the academic realm, as we discussed in Chap. 5. While it has also been successfully used in practice, many existing techniques cannot be applied "out of the box". Their application is obstructed by a certain number of problems that we are going to analyze in detail in this chapter.

In Chap. 9, we look beyond these problems and sketch a vision for an integrated process mining approach in small and medium sized enterprises. Chapter 10 and onward (Parts III and IV) are devoted to solve the problems and realizing this vision.

8.1 Typical Deploy Scenarios

Before analysing in detail the process mining problems, let us present a framework to attribute companies a characterization based on the concept of process awareness. We think it is possible to analyse two different axes: the first measures the *process awareness of the company*; the second measures the *process awareness of the information systems* used within the company. We may define a company as *process aware* when there is a shared knowledge of business process management among the people of the company, who can think and act by processes. This does not necessarily imply that the information systems adopted consciously support processes. That's why, in the second axis, we measure the extent to which the information systems are process aware.

Figure 8.1 proposes four companies, at the "extreme positions". It is worthwhile to note that each of these companies may benefit from process mining techniques. For example, if *Company 1* or *Company 2* decide to move their organizations towards more structured and process oriented businesses, control-flow discovery approaches are extremely valuable. *Company 3*, on the other side, already has a mature infrastructure: in this case it might be interesting to evaluate the performances of the company in order to find possible improvements on the business processes. Finally, *Company 4* can benefit from process mining techniques in several ways: since the information systems adopted do not "force" any process, it might be useful to compare the "ideal

© Springer International Publishing Switzerland 2015 59
A. Burattin: *Process Mining Techniques in Business Environments*, LNBIP 207,
DOI 10.1007/978-3-319-17482-2_8

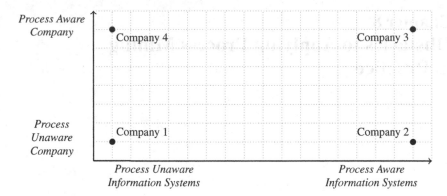

Fig. 8.1 Possible characterization of companies according to their process awareness and to the process awareness of the information systems they use.

processes" with the actual executions, or to evaluate the current performance in order to improve the quality of the executed processes.

In all the scenarios just analysed, when dealing with real-world business process, there are a number of problems that may occur. We have identified three possible "moments of problems":

1. **before** starting the mining phase, when the data have to be prepared;
2. **during** the actual process mining executions;
3. **after** the mining, during the interpretation of the results.

Each of the above-mentioned phases will be analyzed separately, in the three following subsections.

8.2 Problems with Data Preparation

One of the first problems that must be solved is entirely technological. Specifically, it involves the interoperability between data: the log files produced by the information systems must be processed by process mining tools. A well-known process mining software platform is ProM [163, 173, 179]: it is used by researchers for prototyping new algorithms. This software can receive input in two formats: MXML [163] or the recent OpenXES [75] (both are XML-based and easy to read and understand). The main difference between the two formalisms lies on the "extendibility" of the log: OpenXES allows the definition of custom extensions, useful for representing decorative attributes of the log, while MXML does not. Eventually, the interoperability problem (concerning the ProM tool) has been solved with the implementation of the ProM Import Framework [73]. This tool supports the definition of extensions (by adding some plugins) that convert custom log files into MXML or OpenXES.

Another possible problem that may occur is related to the data recorded by Information Systems. Let's consider a typical industrial scenario in which Information Systems are used only to support the process executions. Most of all, these systems are not managing the entire process; instead, they are used only to handle some activities. A typical example is Document Management System (DMS): with such software it is possible to share documents, notify authors about modifications, download the latest version of a document and so on. In a common scenario, DMSs are required to be very flexible, in order to allow the possible *ad hoc* solutions (based on the single case). All the actions executed by the system are recorded in some log files, however many times DMSs are "process unaware" (or process-agnostic) and so are their logs: for example, there is no information on which process or process instance the current action is referring to, even if the system is "describing" a real business process. This problem can be summarized as the problem of applying process mining starting from logs generated by process unaware sources. Specifically, this problem belongs to **P-01** (as presented in Sect. 1.2) and will be named "case ID selection".

The last problem is the presence of some "noise" inside the log. Actually, this issue does not belong to this phase only, but it also spans in the next one. As presented by Christian Günter, in his Ph.D. thesis [72], it is possible to identify several "types of noise". However, independently from the actual possible observations of noise in the log, a log is said to be *noisy* [23] if either:

1. some recorded activities do not match with the "expected" ones, i.e. there exist records of performed activities which are unknown or which are not expected to be performed within the business process under analysis (for example, an activity that is required in the real environment, but that is unknown to the designer);
2. some recorded activities do not match with those actually performed, i.e. activity A is performed, but instead of generating a record for activity A, a record for activity B is generated; this error may be due to a bug introduced into the logging system or to the unreliability of the transmission channel (e.g. a log written to a remote place);
3. the order in which activities are recorded may not always coincide with the order in which the activities are actually performed; this may be due to the introduction of a delay in recording the beginning/conclusion of the activity (e.g. if the activity is a manual activity and the worker delays the time to record the start/end of the activity) or to delays introduced by the transmission channel used to communicate the start/end of a remote activity.

As one can notice, points 2 and 3 of the previous list represent problems related to the "preparation" of the log, i.e. problems that occur before the real mining takes place. These problems can hardly be solved, because information required for the solution is not available, and it cannot be extracted or reconstructed. More generally, these problems are located into a level that is out of the scope of process mining (in particular, this information should be already available), so it seems very hard to correctly reconstruct the missing information. This problem will be addressed in Chap. 10.

8.3 Problems During the Mining Phase

With the term "mining phase" we refer to all the activities that occur between the moment the log is ready for the mining and the final step, that involves the interpretation of results.

Some of the problems that emerge at this time have already been mentioned: there is the noise issue, so it may happen that sometimes the process extracted from the mining phase is not what the process analyst is expecting (i.e. there can be activities that were not supposed to occur, or occurring in the wrong position).

Another issue, related to this phase, is the problem of process unaware sources: consider again the example of a DMS. In that case, it is possible that the generated log could not be as informative as one would expect. For example, activities may be described just with the document names handled by the specific activities, and some details may be missing, like the case identifier. Another problem may occur when considering the opposite scenario: there are too many details about the process that is going to be analysed. In this case, the process mining algorithm has to choose the correct "granularity" for generating the process model, but it has to take advantage of all the available information. For example, in [24], any activity spans over time intervals so, the mining algorithm can exploit this fact in order to extract a better (in the sense of more precise) model. A similar problem occurs when data referring to several *perspectives* is available: it would be desirable to embed all of them into the same representation [29].

The last problem, related to the current phase, is the difficulty in configuring the process mining algorithm: in order to be as general-purpose as possible, such algorithms provide several parameters to be set. For example, the Heuristics Miner algorithm [167] (that will be described in Sects. 5.1 and 11.1) requires thresholds that are useful for the identification of the "noise" (only behaviors that generate values greater than a threshold are considered as genuine). Configuring these parameters is not straightforward, especially for a non-expert user[1]. The actual problem is that, typically, the more "powerful" an algorithm is, the more parameters it requires.

All these problems are analyzed, and some possible solutions are proposed in Chaps. 11, 12, 13 and 14.

8.4 Problems with the Interpretation of the Mining Results and Extension of Processes

The last type of possible problems related to deploy emerges when mined process models are obtained. In this case, there are two issues to tackle.

The first problem is related to the "evaluation" of the mined process: how can we define a rigorous and standard procedure to evaluate the quality of the output

---

[1] Please note that with the term "non-expert user" we identify a user with no experience in process mining, but with notions in business process modeling.

generated by a process mining algorithm? It is possible to compute the performance of the process mining result by comparing the mined model with the original logs, and then observe the discrepancies between what is supposed to happen (the mined model) and what actually takes place (the log). Another important advantage of process mining and, in particular, of control-flow discovery, is acknowledged when a company decides to begin a new business approach, based on Model Driven Engineering [87]: in this case, instead of starting from a new model, it is possible to use the actual set of activities as they are performed in reality. In this case, it is very important to bind activities with roles and originators, so to immediately have an idea, as clear as possible, of the current situation.

The second issue concerns the representation of the model: the risk of creating a graphical representation that is dense of data (e.g. activities, originators, frequencies, dependencies, inputs, outputs, ...) is that it will be hard to be understood and, under certain conditions, useless (mainly because of its cognitive load [107]). The aim is to find the "best" balance between the information presented in the model and the difficulty in reading the model itself. A possible way to deal with this problem is using an interactive approach, where different views can be "enabled" or "disabled" according to the user needs.

These problems will be addressed in Chap. 15.

8.5 Incremental and Online Process Mining

One of the main aims of process mining is control-flow discovery, i.e., learning process models from example traces recorded in event logs. Many different control-flow discovery algorithms have been proposed in the past. Basically, all such algorithms have been defined for batch processing, i.e., a complete event log containing all executed activities is supposed to be available at the moment of execution of the mining algorithm. Nowadays, however, the information systems supporting business processes are able to produce a huge amount of events thus creating new opportunities and interesting challenges from a computational point of view. In fact, when we deal with streaming data, it may be impossible to store all events. Moreover, even if one is able to store all event data, it is often impossible to process them due to the exponential nature of most algorithms. In addition to that, a business process may evolve over time. Manyika et al. [104] report possible ways for exploiting large amount of data to improve the company business. In their paper, *stream processing* is defined as "*technologies designed to process large real-time streams of event data*" and one of the example applications is *process monitoring*.

The challenge to deal with streaming event data is also discussed in the process mining Manifesto[2] [83]. Chapter 17 of this book addresses this problem.

---

[2]The Process Mining Manifesto is authored by the IEEE Task Force on Process Mining (http://www.win.tue.nl/ieeetfpm/).

Chapter 9
Long-term View Scenario

The problems discussed in Chap. 8 refer to various obstacles faced around data availability, quality, and analysis within process mining. In addition to these rather technical obstacles, process mining also has to face organizational challenges: a company and the people involved have to be able to handle and pursue a process mining project.

In this context, the issues are to *integrate* and *embed* the various technical steps into the "organization mechanism". This represents a challenge that small and medium sized enterprises often lack the resources for. In the following chapter, we discuss the related problems and sketch a suitable architecture.

9.1 A Target Scenario

In a long-term view, a possible architecture for exploitation of process mining results is presented in Fig. 9.1. The final aim of this illustrative architecture is to move Small and Medium Enterprises (SME)[1] towards the adoption of "process-centric information systems". According to the latest available statistics [58], SMEs, in the European Union, manage most of the business and this motivates the strong impact that process mining approaches might have in the European market.

Since this long-term view makes sense only if it is applied in real world contexts, it is necessary to solve the problems presented in Sect. 1.2, in order to apply process mining techniques.

Business processes are cross-functional and involve multiple actors and heterogeneous data. Such complexity calls for a structured and planned approach, requiring a substantial amount of competence and human resources. Unfortunately, SMEs typically do not have sufficient resources to invest in this effort. Moreover, many times,

---

[1] According to the "Recommendation concerning the definition of micro, small and medium-sized enterprises" [57], a SME "*is made up of enterprises which employ fewer than 250 persons and which have an annual turnover not exceeding EUR 50 million, and/or an annual balance sheet total not exceeding EUR 43 million.*"

© Springer International Publishing Switzerland 2015   65
A. Burattin: *Process Mining Techniques in Business Environments*, LNBIP 207,
DOI 10.1007/978-3-319-17482-2_9

SMEs do not own a clear and formal description of their business processes since such knowledge is only "mentally held" by few key staff members. It is sufficient that one of these persons quits the company to create difficulties in the maintenance of business processes.

A way to cope with these difficulties is to resort, as far as possible, to formal models for representing business processes within a model-driven engineering (MDE) approach. The rigorous adoption of a MDE approach can lead to an improvement of the productivity in the context of small and medium-sized companies: it can facilitate the quick adaptation of the processes to the changes of the market and to better face variations in the demand for resources. A fundamental issue, however, is how to get useful models, i.e., models that represent relevant and reliable behavior/information about the business processes. There are, at least, three components of a business process that need to be modeled:

1. artifacts (all the digital documents involved in a business process, e.g. invoices, orders, database entries, …);
2. control-flow (ordering constraints among activities);
3. actors (who is actually going to execute the tasks).

Techniques for the automatic extraction of information from the execution of business processes for each one of these components have already been developed: data mining, process mining, and social mining.

9.2 Discussion

Many SMEs, in the European Union, do not take advantage of Process Aware Information Systems (PAIS) and prefer to just use information systems that are not process aware. In addition, for the support of their activities, many other software are used, such as email, free-text documents, … Of course, most of the times, a business process is actually driving these companies, but such knowledge is not encoded into any specification.

In the example architecture of Fig. 9.1, the idea is to present a system that uses many different data-sources from different departments of a company (e.g. document management systems, mobile devices, …). As presented in the diagram, the architecture is divided into three packages: ETL, process mining and MDE. The first package is responsible for the Extraction, the Transformation and the Loading of the data from different data-sources into a single "data-collector". In particular, this data should be extracted from the sources and then cleaned as much as possible, in order to get a uniform and "verified" version of the data. Once a log is available, it is given as input to the second component: the process miner. Considering an ideal approach, the result of this phase is a global process model, where all the aspects are properly described and a "holistic" merge is applied among all the different perspectives. The process model extracted during the second phase can be used in two possible ways: to perform conformance checking analysis (between the model itself and new logs,

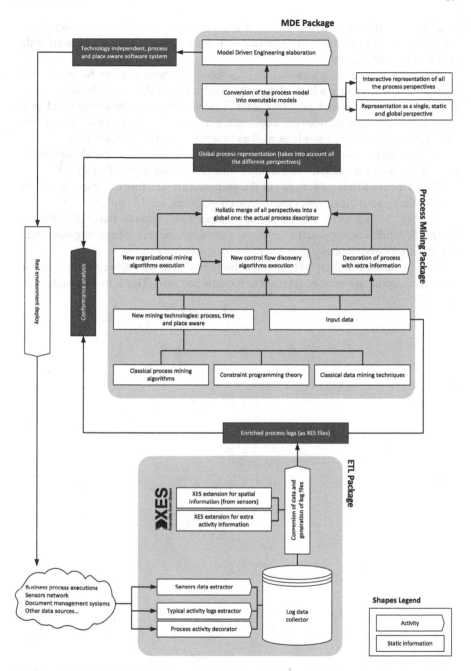

Fig. 9.1 A possible architecture for a global system that spans from the extraction phase to the complete reengineering of the current production process model.

in order to monitor the performances of executions) or as input for a model driven
approach that can automatically generate software or adapt systems to be used in the
production environment.

The impact of the implementation of a similar architecture can be impressive:
it could admit a system which allows the conversion of the actual business into a
process-oriented one, with very low costs. This will increase the ability to adapt to
the requirements arising from the marketplace, in order to speed up the rate at which
SMEs respond to market needs as well as to service or product innovation.

With respect to this book, the "ETL Package" is supposed to deal with data import
and, likely, with incomplete information. This is the main topic of Chap. 10.

The "Process Mining Package" of the proposed architecture encompasses all the
actual mining activities, which are discussed in Chaps. 12, 13 and 14.

Typical "Model Driven Engineering elaborations" include those described in
Chap. 15, which are supposed to evaluate the mined models, comparing them with
the already available ones.

Finally, the entire architecture could be ported into an online settings, but its
internal components must be adapted to the new scenario. This issue is discussed in
Chap. 17.

Part III
Process Mining as an Emerging Technology

This part focuses on problems that arise during the entire process mining phase, as mentioned in Chap. 8: in particular, it is possible to characterize the problems according to the "moments" when they may occur.

The first problem is connected to the preparation of the data to be analyzed by process mining algorithms. The data preparation problem was already introduced in Sect. 8.2. Once data are available, it is possible to start the actual analysis. Different set of problems may arise now, as described in Sect. 8.3. The first problem of the actual mining phase, lies in mining activities using data with a "deep granularity". Specifically, we will consider logs where activities are recorded as time intervals, therefore with a starting and finishing events. The second problem we will face is the complexity in configuring parameters of mining algorithms.We will propose a couple of solutions, both automatic and user-guided. Then, we will analyze how to extend a process model with additional perspectives and how to interpret the mining results. This book part continues by tackling the post-mining problems, as mentioned in Sect. 8.4: two metrics for the evaluation of mining results are proposed. Finally, the last chapter of this part proposes a simple approach to get some data, useful to immediately start testing the capabilities of process mining techniques.

Chapter 10
Data Preparation

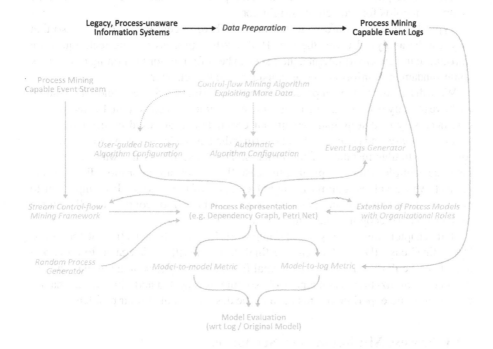

The idea underpinning process mining techniques is that most business processes that are executed with the support of an information system, leave traces of their activity executions and this information, which is stored in the so-called "log files". The aim of process mining is indeed to discover, starting from these logs, as much information as possible. In particular, control-flow discovery aspire to synthesizing a business process model out of data.

In order to perform such reconstructions, it is necessary that the log contains a minimum set of information. With respect to the problems mentioned in Sect. 1.2, this chapter deals with **P-01**, which is also discussed in Sect. 8.2.

© Springer International Publishing Switzerland 2015
A. Burattin: *Process Mining Techniques in Business Environments*, LNBIP 207,
DOI 10.1007/978-3-319-17482-2_10

In particular, all the events, recorded into the log, must provide:

- the name of the activity performed;
- the process case identifier (i.e. a field with the same value for all the executions of the same process instance);
- the execution time of the activity.

This set of information can be considered as the minimum required one. However, beside those, there can be information on:

- the name of the process the activity belongs to;
- the activity originator (if available).

In this context, we consider the name of the process optional because, if it is not provided, it is possible to assume the same process for all the events. The same assumption holds for the activity originator.

Typically, these logs are collected in MXML or, recently, in XES files, so that they can be analyzed using the tool ProM. When process mining techniques are introduced in new environments, the data can be sufficient for the mining or can lack some fundamental information, as considered in this chapter.

We will assume to have a process unaware Information System, and let's assume we have to analyze log data generated from executions of such system. In this context, it is necessary to distinguish two similar concepts that are used in different ways: *process instance* and *case ID*. The first term indicates the logical flow of the activities for one particular instance of the process. The "case ID" is the way the process instance is implemented: typically, the case ID is a set of one or more fields of the dataset, whose values identify a single execution of the process. It is important to note that there can be many possible case IDs although, of course, not all of them may be used to recognize the actual process.

This chapter presents a general framework [30] for the selection of the "most interesting" case IDs, and where the final decision will be delegated to an expert user. In particular, we first preset a formal framework for describing log files and our problem formalization. Then a possible solution is reported and, finally, the chapter ends with some experimental results and the description of similar problems.

10.1 Process Mining in New Scenarios

Consider, for example, Document Management Systems (DMS): they are widely used in large companies, public administrations, municipalities, etc. Clearly, documents can be referred to processes and protocols (consider, for example, the documents involved in supporting the process of selling a product) but they may not contain an explicit reference to it. In the following sections we present our idea, which consists of exploiting the information produced by such support systems, not limiting to DMSs, in order to mine the processes in which the system itself is involved. The nodal point is that these systems typically do not log explicitly the case ID. Therefore, it is necessary to infer this information that enables to relate the system entities (e.g. documents in a DMS) to the process instances.

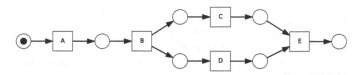

Fig. 10.1 Example of a process model. In this case, activities C and D can be executed in parallel, i.e. in no specific order.

One of the fundamental principles that underpins the idea of "process modeling" is that a defined process can generate any number of concurrent instances (i.e., instances "running" at the same time). Consider, as an example, the process model defined in Fig. 10.1: it is composed of five activities. The process always starts executing A; then B; C and D can be performed in any order and, finally, E is executed. We can observe more than one instance running at the same time: so, for example, at time t we can observe any number of activities executed. Figure 10.2 tries to represent three instances of the same process: the first (c_i) is almost complete; the second (c_{i+1}) is just started and the last one (c_{i+2}) has just completed the first two activities.

In order to identify different instances, it is easy to figure out the need of an element that connects all the observations belonging to the same instance. This element is called "case identifier" (or "case ID"). Figure 10.2 represents it with different colors and with the three labels c_i, c_{i+1} and c_{i+2}.

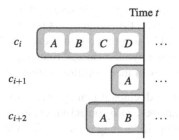

Time	Case ID	Activity
$t-3$	c_i	A
$t-2$	c_i	B
$t-1$	c_i	C
$t-1$	c_{i+2}	A
t	c_i	D
t	c_{i+1}	A
t	c_{i+2}	B

Fig. 10.2 Two representations (one graphical and one tabular) of three instances of the same process (the one of Fig. 10.1).

10.2 Working Framework for Event Logs

The process mining framework we address is based on a set $\mathscr{L}$ of *log entries* originated by auditing activities on a given system. Each log entry $l \in \mathscr{L}$ is a tuple of the form:

$$(activity, timestamp, user, info_1, \ldots, info_m)$$

Table 10.1 An example of log $\mathscr{L}$ extracted from a document management system: all the basic information (such as the activity name, the timestamps and the originator) is shown, together with a set of information on the documents ($info_1 \ldots info_m$). The activity names are: a_1 = "Invoice"; a_2 = "Waybill"; a_3 = "Cash order"; a_4 = "Carrier receipt".

Act.	Timestamp	User	$info_1$	$info_2$	...	$info_m$
a_1	2015-01-02 12:35	Alice	A	2014-06-02	...	
a_2	2015-01-02 12:36	Alice	A	B	...	
a_3	2015-01-03 17:41	Bob	A	2014-06-03	...	
a_4	2015-01-04 09:12	Charlie	A	B	...	
a_1	2015-01-05 08:45	Eve	B	2014-05-12	...	
a_2	2015-01-06 07:21	Alice	B	A	...	
a_3	2015-01-06 11:54	Bob	C	2014-02-20	...	
a_4	2015-01-06 15:15	Charlie	B	A	...	
a_1	2015-01-08 09:55	Bob	D	2014-03-30	...	
a_2	2015-01-08 10:11	Bob	D	C	...	
a_3	2015-01-09 16:01	Bob	C	2014-06-08	...	
a_4	2015-01-09 18:45	Charlie	D	D	...	

In this form, it is possible to identify:

- *activity* the name of the registered activity;
- *timestamp* the temporal instant in which the activity is registered;
- *user* (or *originator*) the agent that executed the registered activity;
- $info_1, \ldots, info_m$ possibly empty additional attributes. The semantics of these additional attributes depend on the activity of the respective log entry. Specifically, given an attribute $info_k$ and activities a_1, a_2, $info_k$, the possible values may represent different information for a_1 and a_2; moreover, the semantics is not explicit. We call these data "decorative" or "additional" since they are not exploited by standard process mining algorithms. Observe that two log entries, referring to the same activity, are not required to share the values of their additional attributes.

Table 10.1 shows an example of such a log in a document management environment. Please note the semantic dependency of attribute $info_2$ on activities: in case of "Invoice" it may represent a date, in case of "Waybill" it may represent the account owner name.

The difference between such log entries and an event in the standard process mining approach is the lack of process instance information. More in general, it can be observed that the source systems we consider do not implement explicitly some workflow concepts, since $\mathscr{L}$ might not come from the sampling of a formalized business process at all.

From now on, we assume a relational algebra point of view over the framework: a log $\mathscr{L}$ is a relation (set of tuples) whose attributes are (*activity*, *timestamp*, *originator*, $info_1, \ldots, info_m$).

As usual, we define the projection operator $\pi_{a_1, \ldots, a_n}$ on a relation R as the restriction of R to attributes $a_1, \ldots, a_n$ (observe that duplicates are removed by projection, otherwise the output would not be a set).

For the sake of brevity, given a set of attributes $A = \{a_1, \ldots, a_n\}$, we denote a projection on all the attributes in A as $\pi_A(R)$. Similarly, given an attribute a, a value constant v and a binary operation θ, we define the selection operator $\sigma_{a\theta v}(R)$ as the selection of tuples of R for which θ holds between a and v; for example, given $a = activity$, $v = $ "Invoice", and θ being the identity function, $\sigma_{a\theta v}(\mathscr{L})$ is the set of elements of $\mathscr{L}$ having "Invoice" as value for attribute $activity$. For a complete survey about relational algebra concepts refer to [53].

It is worthwhile to notice that relational algebra does not deal with tuples ordering, which is on the other hand a crucial issue in process mining. However, this is not an impassable problem since *(i)* the log can be sorted whenever required and *(ii)* we are now concentrating on the generation of a log suitable for applying process mining techniques (and not a process mining algorithm itself).

From now on, identifiers $a_1, \ldots, a_n$ will range over activities. Moreover, given a log $\mathscr{L}$, we define the set $\mathscr{A}(\mathscr{L}) = \pi_{activity}(\mathscr{L})$ (distinct activities occurring in $\mathscr{L}$). Finally, we denote with $\mathscr{I}$ the set of attributes $\{info_1, \ldots, info_m\}$.

As stated above, our efforts are concentrated on extracting flow of information from $\mathscr{L}$, that is, guessing the case ID for each entry $l \in \mathscr{L}$, according to the following restrictions on the framework.

Fixed a log $\mathscr{L}$, we assume that:

1. given a log entry $l \in \mathscr{L}$, if a case ID exists in l, then it is a combination of values in the set of attributes $PI \subseteq \mathscr{I}$ (i.e. *activity, timestamp, originator* do not participate in the process instance definition);
2. given two log entries $l_1, l_2 \in \mathscr{L}$ such that $\pi_{activity}(l_1) = \pi_{activity}(l_2)$, if PI contains the case ID for l_1, then it also contains the case ID for l_2 (i.e., process instance attributes set is fixed per activity); this is implied by the assumption that the semantics of additional fields is a function of the activity, as stated above.

10.3 Identification of Process Instances

From the basis we just defined, one can deduce that the process instance has to be guessed as a subset of $\mathscr{I}$; however, since the semantics is not explicitly given, it cannot be exploited to establish correlation between activities, hence the process instance selection must be carried out looking at the entries $\pi_i(\mathscr{L})$, for each $i \in \mathscr{I}$.

Nonetheless, since the semantics of $\mathscr{I}$ is a function of the activity, the selection should be performed for each activity in $\mathscr{A}(\mathscr{L})$, for each attribute in $\mathscr{I}$, resulting in a computationally expensive procedure. In order to reduce the search space, some intuitive heuristics are depicted: their application resulted successful in our experimental environment.

10.3.1 Exploiting A-priori Knowledge

Experts of the source domain typically hold some knowledge about the data that can be exploited to discard the less promising attributes.

Let a be an activity, and $\mathscr{C}(a) \subseteq \mathscr{I}$ the set of attributes candidate to participate in the process instance definition, with respect to the given activity. Clearly, if no *a-priori* knowledge can be exploited to discard some attributes, then $\mathscr{C}(a) = \mathscr{I}$.

The experiments we carried out helped us define some simple heuristics for reducing the cardinality of $\mathscr{C}(a)$, basing on:

- assumptions on the data type (e.g. discarding timestamps);
- assumptions on the case ID expected format, like average length upper and lower bounds, length variance, presence or absence of given symbols, etc.

It is worthwhile to notice that this procedure may lead to discard all the attributes $info_i$ for some activities in $\mathscr{A}(\mathscr{L})$. In the following formula we denote with $\mathscr{A}(\mathscr{C})$ the set of all the activities that overcome this step, that is

$$\mathscr{A}(\mathscr{C}) = \bigcup_{a \in \mathscr{A}(\mathscr{L})} \{a \mid \mathscr{C}(a) \neq \emptyset\}.$$

$\mathscr{A}(\mathscr{C})$ contains all the activities which have some candidate attribute, that is, all the activities that can participate in the process we are looking for.

10.3.2 Selection of the Identifier

After the search space has been reduced and the set $\mathscr{C}(a)$ has been computed for each activity $a \in \mathscr{A}(\mathscr{L})$, we must select those elements of $\mathscr{C}(a)$ that participate in the process instance. The only information we can exploit in order to automatically perform this selection is the amount of data shared by different attributes.

Aiming at modeling real settings, we fix a sharing threshold T, and we retain as candidate those subsets of $\mathscr{C}(a)$ that share at least T entries with some attribute sets of other activities. This threshold must be defined with respect to the number of distinct entries of the involved activities, for instance as a fraction of the number of entries of the less frequent one.

Let (a_1, a_2) be a pair of activities, such that $a_1 \neq a_2$ and let PI_{a_1} and PI_{a_2} the corresponding process instances field. We define the function S that calculates the shared values among them:

$$S(a_1, a_2, PI_{a_1}, PI_{a_2}) = \left| \pi_{PI_{a_1}}(\sigma_{activity=a_1}(\mathscr{L})) \bigcap \pi_{PI_{a_2}}(\sigma_{activity=a_2}(\mathscr{L})) \right|$$

Observe that, in order to perform the intersection, it must hold $|PI_{a_1}| = |PI_{a_2}|$. Using such function, we define the process instance candidates for (a_1, a_2) as:

$$\varphi(a_1, a_2) = \left\{ (PI_{a_1} \in \mathscr{P}(\mathscr{C}(a_1)), PI_{a_2} \in \mathscr{P}(\mathscr{C}(a_2))) \mid S(a_1, a_2, PI_{a_1}, PI_{a_2}) > T \right\}$$

where $\mathscr{P}$ denotes the power set.

Elements of $\varphi(a_1, a_2)$ are pairs, whose components are those attribute sets, respectively of a_1, a_2, that share a number of values greater than T (i.e. the cardinality of the intersection of PI_{a_1} and PI_{a_2} is greater than T). In the following, we denote with φ_a the set of all the candidate attribute sets for activity a, i.e.:

$$\varphi_a = \left\{ PI \mid \exists a_1 \in \mathscr{A}(\mathscr{C}), PI_{a_1} \in \mathscr{P}(\mathscr{C}(a_1)).(PI, PI_{a_1}) \in \varphi(a, a_1) \right\}.$$

This formula figures out some candidate process instances that may relate two activities: it is worthwhile to note, however, that our target is the correlation of a set of activities whose cardinality is in general greater than 2. Actually, we want to build a sequence $S = a_{S_1}, \ldots, a_{S_n}$ of distinct activities. Nonetheless, given activity a_{S_i}, there may be a number of choices for $a_{S_{i+1}}$, and then a number of process instances in $\varphi(a_{S_i}, a_{S_{i+1}})$. Hence, a number of sequences may be built.

We call *chain* a finite sequence C of n components of the form $[a, X]$, being a an activity and $X \in \varphi_a$. We can denote it as follows:

$$C = \left[a_1, PI_{a_1}\right], \left[a_2, PI_{a_2}\right], \ldots, \left[a_n, PI_{a_n}\right]$$

such that $\left(PI_{a_i}, PI_{a_{i+1}}\right) \in \varphi(a_i, a_{i+1})$, with $i \in [1, n-1]$. We denote with C_i^a the i-th activity of the chain C, and with C_i^{PI} its i-th PI.

Observe that a given activity must appear only once in a chain, since a process instance is defined by a single attribute set. Given a chain C ending with element $[a_j, PI_{a_j}]$, we say that C is *extensible* if there exists an activity $a_k \in \mathscr{A}(\mathscr{C})$ such that $(PI_{a_j}, X) \in \varphi(a_j, a_k)$, for some set $X \in \mathscr{P}(\mathscr{C} a_k))$. Otherwise, C is said to be *complete*. Moreover, we say that an activity a occurs in a chain C, denoted $a \in C$, if there exists a chain component $[a, X]$ in C for some attribute set X. Since an activity can occur in more than one chain with different process instances, in some case we write $PI_{a_i, C}$ to denote the process instance of activity a_i in chain C. Finally, let $\mathscr{A}(C)$ denote the set of activities occurring in a chain C. The empty chain is denoted with $\bot$.

Given a chain C, we define the average value sharing $S(C)$ among selected attributes of C as:

$$S(C) = \frac{\sum_{1 \leq i < n} S\left(C_i^a, C_{i+1}^a, C_i^{PI}, C_{i+1}^{PI}\right)}{n - 1}$$

where n denotes the chain length.

All the possible complete chains on $\mathscr{L}$ are built according to Algorithms 1 and 2.

Algorithm 1 calls Algorithm 2 for all the extensible chains of length 1. Observe that the pseudo code of Algorithms 1 and 2 builds also some chains which are permutations of one another.

10.3.2.1 Match Heuristics

In computing the amount of data shared by two activities via function φ, heuristics approaches may help in modeling the complexity of a real domain. Actually, the

---

Algorithm 1. Build Chains

---

1 **foreach** $a \in \mathcal{A}(\mathcal{C})$ **do**
2   **foreach** $PI \in \varphi_a$ **do**
3 | Extend Chain($[a, PI]$)   /* see Algorithm 2 */
4   **end**
5 **end**

Algorithm 2. Extend Chain

Input: a chain $C = [a_1, PI_1], \ldots, [a_{i-1}, PI_{i-1}]$
1 **foreach** $a_i \in \mathcal{A}(\mathcal{C}) \mid a_i \notin C$ **do**
2   **foreach** $PI_i \in \varphi_{a_i} \mid (PI_{i-1}, PI_i) \in \varphi(a_{i-1}, a_i)$ **do**
3 | $C = C, [a_i, PI_i]$
4 | **return** *Extend Chain*(C)   /* recursive call */
5   **end**
6 **end**

---

comparison performed between values does not need to be an identity match; instead, a fuzzy match can be implemented. Guided by this basic heuristics, we can substitute the intersection operator in φ with an approximation of it, whose definition may be domain specific or not. Simple examples we tested in our experimental environment are:

- equality match up to X leading characters,
- equality match up to Y trailing characters,

and their combinations. In general it is possible to use a measure for string distance.

10.3.3 Results Organization and Filtering

In the previous sections we have shown how to compute a number of chains (i.e., a number of logs); in general, a domain expert is able to discriminate between "good chains" and less reasonable ones, but this could be a demanding task. Here we present the problem of comparing different chains: in order to address this issue, it is worthwhile to analyze a methodology that helps restricting the number of possible chains.

Generally, we reject a chain in favor of another one if and only if the latter contains all the activities of the former, and it is either simpler or it supports a higher confidence. Example of parameters taken into account might be:

- the number of attributes in the process instance of a chain component (recall that each component has the same number of process instance attributes): a chain that concerns less attributes may be considered simpler, thus preferable since more readable for a human analyst;

- the cardinality of the shared value between chain components ($S(\cdot)$): a chain whose share factor is higher, gives higher confidence. This parameter could be tuned by a threshold.

Let $\mathscr{H}$ be the set of complete chains computed by Algorithms 1 and 2, without permutations. Given two chains C_1 and C_2:

$$C_1 = \big[a_1, PI_{a_1}\big], \ldots, \big[a_n, PI_{a_n}\big]$$
$$C_2 = \big[b_1, PI_{b_1}\big], \ldots, \big[b_m, PI_{b_m}\big]$$

in the set $\mathscr{H}$, we define an ordering operator $\sqsubseteq$ as:

$$A \sqsubseteq B \Leftrightarrow \begin{cases} \big|A_1^{PI}\big| \geq \big|B_1^{PI}\big| & \text{if } \mathscr{A}(A) = \mathscr{A}(B) \wedge S(A) = S(B) \\ S(A) \leq S(B) & \text{if } \mathscr{A}(A) = \mathscr{A}(B) \wedge S(A) \neq S(B) \\ \mathscr{A}(A) \subseteq \mathscr{A}(B) & \text{otherwise} \end{cases}$$

The operator $\sqsubseteq$ defines a reflexive, antisymmetric, and transitive relation over chains, hence $(\mathscr{H}, \sqsubseteq)$ is a partially ordered set [137]. For the sake of simplicity, in the above formulation we do not use any threshold to tune the value sharing comparison.

The ordering we defined, strives to equip the framework with a notion of "best chains", i.e., those chains which could be suggested to a domain expert.

10.3.4 Deriving a Log to Mine

For each chain C with positive length, we can build a log $\mathscr{L}'$ whose tuples have the form:

$$(activity, timestamp, user, case\ ID, processID)$$

Please observe that the process instance we selected is a set of attributes, whereas a single one is expected by standard process mining techniques. Hence, a composition function k from a set of values to a single one is needed (a straightforward example of k is string concatenation). The log $\mathscr{L}'$ is obtained, starting from $\mathscr{L}$s with the execution of Algorithm 3.

Observe that the order on the chain components does not influence the process instance selection. For this reason, in order to build the log $\mathscr{L}'$, once all the chains are complete (no more extensible), it is possible to ignore the chains that are permutations of a given one. Thus, some chains computed by Algorithm 1 can be discarded.

It is interesting to observe, however, that maximal elements in the poset represent different processes. A conservative approach compels us considering each maximal chain as defining a distinct process. The following example illustrates the reason why we chose this approach. Given two maximal chains C_1 and C_2:

Algorithm 3. Conversion of $\mathscr{L}$ to $\mathscr{L}'$

   Input: $\mathscr{H}$: set of chains; k: case ID composition function

1 $\mathscr{L}' \leftarrow \emptyset$
2 $chainNo \leftarrow 0$
3 **foreach** $C \in \mathscr{H}$ **do**
4     $\mathscr{L}_C \leftarrow \sigma_{activity \in \mathscr{A}(C)}(\mathscr{L})$
5   **foreach** $l \in \mathscr{L}_C$ **do**
6       $activity \leftarrow \pi_{activity}(l)$
7       $timestamp \leftarrow \pi_{timestamp}(l)$
8       $originator \leftarrow \pi_{originator}(l)$
9       $caseid \leftarrow k\left(\pi_{PI_{activity,C}}(l)\right)$
10      $processid \leftarrow chainNo$
11      $\mathscr{L}' \leftarrow \mathscr{L}' \cup \{(activity, timestamp, originator, caseid, processid)\}$
12   **end**
13    $chainNo \leftarrow chainNo + 1$
14 **end**
15 **return** $\mathscr{L}'$

$$C_1 = \ldots, \left[a_{i-1}, PI_{a_{i-1}}\right], \left[a_i, PI_{a_i}\right], \left[a_{i+1}, PI_{a_{i+1}}\right], \ldots$$
$$C_2 = \ldots, \left[b_{j-1}, PI_{b_{j-1}}\right], \left[b_j, PI_{b_j}\right], \left[b_{j+1}, PI_{b_{j+1}}\right], \ldots$$

where $PI_{a_{i-1}} \neq PI_{b_{j-1}}$; $PI_{a_i} \neq PI_{b_j}$; $PI_{a_{i+1}} \neq PI_{b_{j+1}}$ and $a_i = b_j$. In other words, C_1 and C_2 contain the same activity a_i but with different process instances. Considering C_1 and C_2 as belonging to the same process is not desirable, since this can lead to inconsistent control-flow reconstruction. Hence, each maximal chain defines a process and the domain expert is in charge of recognizing if different chains belong to the same real process. During the conversion of $\mathscr{L}$ to the process log $\mathscr{L}'$, we assign as process ID a chain counter.

10.4 Experimental Results

As explained, the problem of case ID identification is common to many businesses. In particular, we tested our procedure in data coming from the company Siav S.p.A.[1]. In this case, the existing implementation is limited to process instances constituted by a single attribute (e.g., $|PI_i| = 1$), due both to *a-priori* knowledge about the domain and computational requirements. In particular, all the pre-processing steps that reduce the search space have been implemented as Oracle store procedures, written in PL/SQL. Then the chain building algorithms are implemented in C#. Moreover, for improving performances, we do not compute the heuristics on the whole log, but on a fraction of random entries.

---

[1]*"Siav is a software development and IT services company specialized in electronic documents management. It is an industry specialist in Italy with over 250 people employed and around 3000 installations".* From http://www.siav.com/.

Table 10.2 Results summary. Horizontal lines separate different log sources (datasets). The table also shows the total number of chains, the maximal chains and the chains pointed out by the expert.

$|\mathscr{L}'|$	$|\mathscr{A}(\mathscr{L}')|$	$|\mathscr{I}|$	Time	$|\mathscr{H}|$	Max. ch.	Exp. ch.
10 000	13	26	6 s	3	2	1
20 000	39	26	20 s	5	2	1
40 000	47	26	1 m 40 s	8	3	2
60 000	2	18	2 s	2	1	0
140 000	4	18	15 s	3	1	1
20 000	12	16	40 s	3	3	1
30 000	16	16	2 m	11	3	1

We tried our implementation on logs coming from a document management system; the source log is reduced to the form described in Sect. 10.2 after undergoing some preprocessing steps.

We applied the algorithms to real logs, obtaining concrete results, validated by domain experts. Table 10.2 summarizes the main information: please note that the expert chains are always within the set of maximal chains (computed by the algorithm), since they selected among the firsts. Figure 10.3 shows how chains evolve when the log cardinality scales up: in particular, notice that the number of chains tends to increase, while the poset structure tears down the number of chains we presented to the domain experts. Figure 10.4 plots the processing time: it is a function of the log cardinality, of the number of activities in the log (i.e. the number of possible chain components), and of the number of decorative attributes (i.e. the number of possible ways of chaining two components).

The knowledge of the application domain gave us the opportunity to implement some heuristics, as explained in Subsect. 10.3.1. The following criteria were selected in order to reduce the search space:

- a candidate attribute must have a string type (i.e., we discard timestamps and numeric types, that in our case mostly represent prices);
- the values of a candidate attribute must fulfill these requirements:

  - maximum average length: 20 characters,
  - minimum average length: 3 characters,
  - maximum variation with respect to the average length: 10.

Finally, we relaxed the intersection operator in φ requiring values identity up to the first leading character and up to 2 trailing characters.

The experiments were carried out on an Intel Core2 Quad at 2,4 GHz, equipped with 4 GB of RAM. The DBMS where the logs were stored was local to the machine, thus no network overhead has to be considered.

10.5 Similar Problems and Solutions

The problem of relating a set of activities to the same process instance is already known in literature. In [59], Ferreira and Gillblad presented an approach for the identification of the case ID based on the Expectation-Maximization (EM) technique.

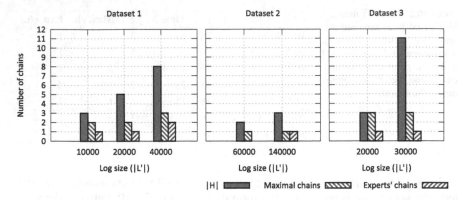

Fig. 10.3 This figure plots the total number of chains identified, the number of maximal chains and the number of chains the expert will select, given the size of the preprocessed log.

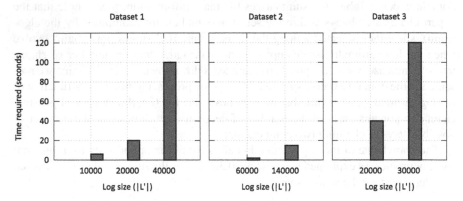

Fig. 10.4 This figure represents the time (expressed in seconds) required to extraction chains, given the size of the preprocessed log.

The most positive characteristic of this approach lies in its generality, which allows its execution in all possible scenarios. However, it suffers of two drawbacks: its computational complexity; and problems deriving from reaching the local optimum of the likelihood function.

Other approaches, such as the one presented by Ingvaldsen et al., in [55] and in [56], use the input and output produced by the activities registered in the SAP ERP. In particular, they construct "process chains" (composed by activities belonging to the same instance) by identifying the events that produce and consume the same set of resources. The assumption underpinning this approach (i.e., the presence of resources produced and consumed by two "joint" activities) seems too strong for a broad and general application.

Other works that deal with the same issue are presented in [60, 180], but all of them solve only partially the problem, or impose specific constraints on the environment, clearly limiting the applicability in general scenarios. In [116], authors present a detailed review of the literature on this field and describe a novel approach that allows the collection of data to be mined. The technique described in this work, however, requires the modification of the source code of the application, but this is not always feasible. An empirical study on an industrial application is provided as well.

The most important difference between our approach and others in literature is that, in our case, the information on the process instance is "hidden inside the log" (we do now know which are the fields with the required information), and therefore it has to be extracted properly. Such a difference is very important for two fundamental reasons:

1. the settings we required are sufficiently general to be observed in a number of real environments;
2. our technique is devised for this particular problem, hence can be more efficient than others.

10.6 Summary

This chapter presents an approach for the identification of process instances on logs generated from systems that are not process-aware.

Process instance information is guessed using additional meta-data, typically available when dealing with software systems, with respect to a standard process mining framework.

The described procedure is entirely based on the information that decorates documents (this work is a generalization of a real business case related to a document management system, where discovering the process instance means correlating different document set), and relies on a relational algebra approach. Moreover, we deem that our generalization can be fairly adoptable in a number of domains, with a reasonable effort.

With respect to the problems pointed out in Sect. 1.2, this chapter deals with **P-01**: problems occurring *before* the actual mining, with the data preparation (see Sect. 8.2). In general, the provided approach solves the problems, but requires the interaction of a domain expert.

Chapter 11
Heuristics Miner for Time Interval

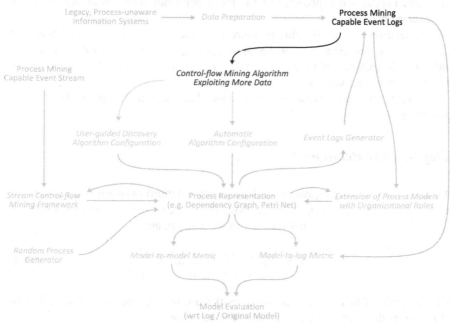

Many control-flow discovery algorithms proposed up to now, assume that each activity is considered instantaneous. This is due to the fact that usually a single log for each performed activity is recorded, regardless of the duration of the activity. In many practical cases, however, activities involve a span of time, so they can be described by time intervals (couples of time points). Of course, not recording the duration of activities makes mining quite hard. Since in some cases information about duration of some activities is available, it is wise to use such information.

For the above-given reasons, the generalization proposed in this section allows the treatment of time intervals. Exploiting this information, a "better" (i.e. closer to the model that originated the logs) process model can be mined, without modifying

© Springer International Publishing Switzerland 2015
A. Burattin: *Process Mining Techniques in Business Environments*, LNBIP 207,
DOI 10.1007/978-3-319-17482-2_11

the overall complexity of the original algorithm and, in addition, preserving backward compatibility.

This chapter presents the generalization of a popular process mining algorithm, named Heuristics Miner, to time intervals. In particular, it will be shown that the possibility to use time interval information, when available, allows the algorithm to produce better workflow models.

With respect to the problems mentioned in Sect. 1.2, this chapter deals with **P-02**, which is also discussed in Sect. 8.3.

11.1 Heuristics Miner

Heuristics Miner, already briefly presented in Sect. 5.1, is a process mining algorithm that uses a statistical approach to mine the dependency relations among activities represented by logs.

The relation $a >_W b$ holds iff there is a trace $\sigma = \langle t_1, t_2, \ldots, t_n \rangle$ and $i \in \{1, \ldots, n-1\}$ such that $\sigma \in W$ and $t_i = a$ and $t_{i+1} = b$. The notation $|a >_W b|$ indicates the number of times that, in W, $a >_W b$ holds (no. of times activity b directly follows activity a).

The next subsections presents a detailed list of all the formulae required by Heuristics Miner.

Dependency Relations ($\Rightarrow$)

An edge (that usually represents a dependency relation) between two activities is added if its *dependency measure* is above the value of the *dependency threshold*. This relation is calculated, between activities a and b, as:

$$a \Rightarrow_W b = \frac{|a >_W b| - |b >_W a|}{|a >_W b| + |b >_W a| + 1} \tag{11.1}$$

The rationale of this rule is that two activities are in a dependency relation if most of the times they are in the specifically required order.

AND/XOR Relations ($\wedge$, $\otimes$)

When an activity has more than one outgoing edge, the algorithm has to decide whether the outgoing edges are in AND or XOR relation (i.e. the "type of split"). Specifically, it has to calculate the following quantity:

$$a \Rightarrow_W (b \wedge c) = \frac{|b >_W c| + |c >_W b|}{|a >_W b| + |a >_W c| + 1} \tag{11.2}$$

If this quantity is above a given *AND threshold*, the split is an AND-split, otherwise it is considered to be in XOR relation. The rationale, in this case, is that two activities are in an AND relation if most of the times they are observed in no specific order (so one before the other and *vice versa*).

Long Distance Relations ($\Rightarrow^l$)

Two activities a and b are in a "long distance relation" if there is a dependency between them, but they are not in direct succession. This relation is expressed by the formula:

$$a \Rightarrow^l_W b = \frac{|a \ggg_W b|}{|b| + 1} \tag{11.3}$$

where $|a \ggg_W b|$ indicates the number of times that a is directly or indirectly followed by b in the log W (i.e. if there are other different activities between a and b). If this formula value is above a *long distance threshold*, then a long distance relation is added into the model.

Loops of Length One and Two

A loop of length one (i.e. a self loop on the same activity) is introduced if the quantity:

$$a \Rightarrow_W a = \frac{|a >_W a|}{|a >_W a| + 1} \tag{11.4}$$

is above a *length-one loop threshold*. A loop of length two is considered differently: it is introduced if the quantity:

$$a \Rightarrow^2_W b = \frac{|a >^2_W b| + |b >^2_W a|}{|a >^2_W b| + |b >^2_W a| + 1} \tag{11.5}$$

is above a *length-two loop threshold*. In this case, the $a >^2_W b$ relation is observed when a is directly followed by b and then there is a again (i.e. there exists a trace $\sigma = \langle t_1, t_2, \ldots, t_n \rangle$ and $i \in \{1, \ldots, n-2\}$ such that $\sigma \in W$ and $t_i = a$ and $t_{i+1} = b$ and $t_{i+2} = a$).

Running Example

Let's consider the process model and the log of Fig. 11.1. Given the set of activities $\{A, B_1, B_2, C, D\}$, a possible log W, with 10 process instances, is presented in Fig. 11.1(a) (please note that the notation $\langle \cdots \rangle^n$ indicates n repetitions of the same sequence). Such log can be generated starting from executions of the process model of Fig. 11.1(b).

$$W = \{\langle A, B_1, B_2, C, D \rangle^5 ; \langle A, B_2, B_1, C, D \rangle^5\}$$

(a) Example of process log W with 10 process instances (^{n} indicates n repetitions of the same sequence).

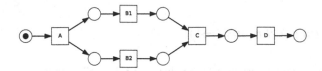

(b) Example of a possible process model that generates the log W.

Fig. 11.1 Example of a process model and a log that can be generated by the process.

In the case reported in figure, the main measure (dependency relation) builds the following relation:

$$
\begin{array}{c c c c c c}
 & A & B_1 & B_2 & C & D \\
A & 0 & 0.8\overline{3} & 0.8\overline{3} & 0 & 0 \\
B_1 & -0.8\overline{3} & 0 & 0 & 0.8\overline{3} & 0 \\
B_2 & -0.8\overline{3} & 0 & 0 & 0.8\overline{3} & 0 \\
C & 0 & -0.8\overline{3} & -0.8\overline{3} & 0 & 0.9\overline{09} \\
D & 0 & 0 & 0 & -0.9\overline{09} & 0
\end{array}
$$

Starting from this relation and considering – for example – a value 0.8 for the *dependency threshold*, it is possible to identify the split from activity A to B_1 and B_2. In order to identify the type of the split, it is necessary to use the AND measure (Eq. 11.2):

$$A \Rightarrow_W (B_1 \wedge B_2) = \frac{5+5}{5+5+1} = 0.9\overline{09}$$

So, considering – for example – an *AND-threshold* of 0.9, the type of the split is set to AND.

Common implementations of Heuristics Miner consider, as default value for *dependency threshold*, 0.9, instead for the *AND-threshold* the default value is set to 0.1.

11.2 Activities as Time Interval

Heuristics Miner considers each activity as an instantaneous event, either if each activity lasts for a certain amount of time. In the example shown in Fig. 11.2, regardless of whether the starting or the finishing time is used, the algorithm will mine always a "linear sequence of dependencies" between all activities (so D depends on C, C

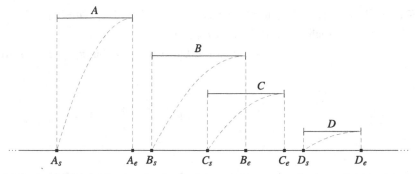

(a) Representation of the time intervals of the activities and projections of the *start* and *end* events on the timeline.

(b) Model that can be mined considering only the *start* event type.

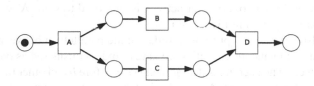

(c) Model that is mined using the intervals and Heuristics Miner++.

Fig. 11.2 Example of model composed by activities with completely different behaviours if mined as time intervals or instantaneous events.

on B and B on A). However, as we can see, there is actually no causal dependency between activities B and C, since they are partially overlapped in time.

In order to extend the algorithm to cope with time intervals, it is necessary to provide a new definition for the direct succession relation in the time intervals context. With an activity represented as a single event, we have that $X >_W Y$ iff $\exists \, \sigma = \langle t_1 \ldots, t_n \rangle$ and $i \in \{1, \ldots, n-1\}$ such that $\sigma \in W$, $t_i = X$ and $t_{i+1} = Y$. This definition has to be modified to cope with activities represented by time intervals.

First of all, given an event e, let's define with $\pi_{activity}(e)$ the activity name the event belongs to, and with $\pi_{type}(e)$ the type of the event (either *start* or *end*).

The new succession relationship $X \overline{>}_W Y$ between two activities is defined as follow:

Definition 11.1 (Direct succession relation, $\overline{>}$). Let a and b be two interval activities (not instantaneous) in a log W, then

$$a \overset{>}{}_W b \text{ iff } \exists \, \sigma = \langle t_1, \ldots, t_n \rangle \text{ and } i \in \{2, \ldots, n-2\}, \, j \in \{3, \ldots, n-1\}$$
$$\text{such that } \sigma \in W, \, t_i = a_{\text{end}} \text{ and } t_j = b_{\text{start}} \text{ and}$$
$$\forall k \text{ such that } i < k < j \text{ we have that } \pi_{type}(t_k) \neq \text{start}.$$

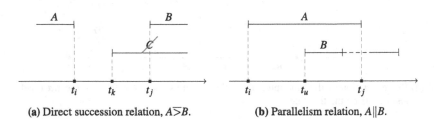

(a) Direct succession relation, $A \overset{>}{} B$.   (b) Parallelism relation, $A \| B$.

Fig. 11.3 Visual representation of the two new definitions introduced by Heuristics Miner++.

Less formally, we can say that two activities, to be in a direct succession relation, must meet the condition for which the termination of the first occurs before the start of the second: between these two, no other activity is supposed to start. A representation of this relation is reported in Fig. 11.3(a).

There is also a new concept to be introduced: the parallelism between two activities. With the instantaneous activities we have a and b considered as parallel when they are observed in no specific order (sometimes a before b and other times b before a), so $(a >_W b) \wedge (b >_W a)$. Actually, this definition may seem odd, but without the notion of "duration", there is no straightforward definition of parallelism.

In the new context, considering not-instantaneous events, the definition of parallelism is easier and more intuitive:

Definition 11.2 (Parallelism relation, $\|$). Let a and b be two interval activities (not instantaneous) in a log W, then

$$a \|_W b \text{ iff } \exists \, \sigma = \langle t_1, \ldots, t_n \rangle \text{ and } i, j, u, v \in \{1, \ldots, n\}$$
$$\text{with } t_i = a_{\text{start}}, t_j = a_{\text{end}} \text{ and } t_u = b_{\text{start}}, t_v = b_{\text{end}}$$
$$\text{such that } \quad u < i < v \quad \vee \quad i < u < j.$$

More intuitively, this definition indicates two activities as parallel if they are overlapped or if one contains the other, as represented in Fig. 11.3(b).

Referring to the notion of "intervals algebra" introduced by Allen [6] and the macro-algebra $A_3 = \{\prec, \cap, \succ\}$, as Golumbic and Shamir described in [67], we can think the direct succession relation as the "preceedings" ($a \prec b$) one and the parallelism relation as the "intersection" ($a \cap b$) one.

We not only modified the notions of relations between two activities, we also improved the algorithm performance modifying the formulae for the statistical dependency and to determine the relation type (AND or XOR).

The new formulation of the dependency measure is:

$$a \Rightarrow_W b = \frac{|a \overset{>}{>}_W b| - |b \overset{>}{>}_W a|}{|a \overset{>}{>}_W b| + |b \overset{>}{>}_W a| + 2|a\|_W b| + 1} \tag{11.6}$$

the new formulation of the AND relation is:

$$a \Rightarrow_W (b \wedge c) = \frac{|b \overset{>}{>}_W c| + |c \overset{>}{>}_W b| + 2|a\|_W b|}{|a \overset{>}{>}_W b| + |a \overset{>}{>}_W c| + 1} \tag{11.7}$$

In this case, the notation $|X\|_W Y|$ refers to the number of times that, in W, activity X and Y are in parallel relation.

In Eq. 11.6, in addition to the usage of the new direct succession relation, we introduced the parallel relation in order to reduce the likelihood to see, in the mined model, the activities in succession relation if in the log they were overlapped.

In the second formula, Eq. 11.7, we inserted the parallelism counter in order to prefer the selection of an AND relation if the two activities are overlapped in the log. In both cases, because of the symmetry of the $\|$ relation, a factor 2 is introduced for parallel relations.

With the new formulae, we obtain "backward compatibility" with the original Heuristics Miner algorithm: if the log does not contain information about interval[1] the behavior is exactly the same of the "standard" Heuristics Miner. This happens because any two activities a and b will never be in parallel relation, i.e. $|a\|_W b| = 0$. We can use this feature to tackle logs with a mixture of activities expressed as time intervals and instantaneous, improving the performances without losing the Heuristics Miner benefits. The new algorithm is called Heuristics Miner++ [24].

11.3 Experimental Results

To assess the efficacy of our new algorithm we performed some tests, on a single, artificial, process; on a series of different processes and logs; and on a real dataset.

First Test: Single Process

The given algorithm has been implemented in the ProM 5.2 process mining framework. In the first test, we tried to generate a random process with six activities. Each of them is composed of a *start* and a *complete* event. The generated log contains 1000 cases and so, in total, 12000 events are recorded. Moreover, 10 % of the traces contain some noise that, in this case, consists of a random swap of two events of the trace. Results of the mining are presented in Fig. 11.4. Figure 11.4(a) proposes the Petri net extracted out of the log using the "classical version" of Heuristics Miner; Fig. 11.4(b) presets the Petri net mined using Heuristics Miner++.

---

[1]In case there are no intervals, it is possible to add "special intervals" to the log, where each activity starts and finishes at the same time.

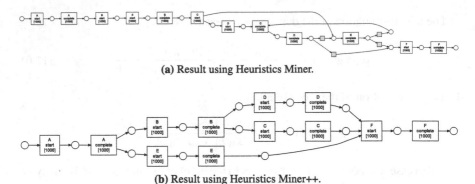

(a) Result using Heuristics Miner.

(b) Result using Heuristics Miner++.

Fig. 11.4 Comparison of mining results with Heuristics Miner and Heuristics Miner++.

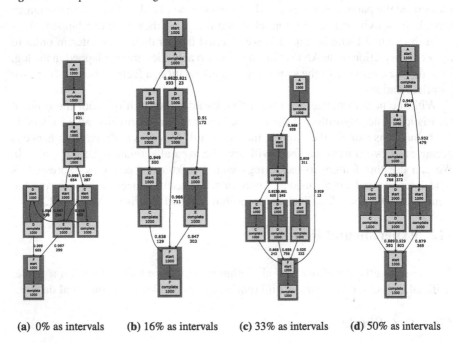

(a) 0% as intervals   **(b)** 16% as intervals   **(c)** 33% as intervals   **(d)** 50% as intervals

Fig. 11.5 Mining results with different percentages of activities (randomly chosen) expressed as time interval. Already with 50 % the "correct" model is produced.

Second Test: Multiple Processes

For the second test, we decided to try our algorithm against a set of different processes, to see the evolution of the behavior when only some traces contain activities as time interval.

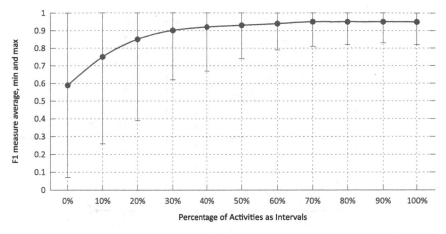

Fig. 11.6 Plot of the F_1 measure averaged over 100 processes logs. Minimum and maximum average values are reported as well.

The dataset we produced contains 100 processes and, for each of them, 10 logs (with 500 instances each) are generated. Considering the 10 logs, the first one contains no activity as time interval; in the second only one activity (randomly different) is expressed as time interval; in the third two of those are intervals and so on, until all activities are expressed as time intervals (Fig. 11.6).

The algorithm Heuristics Miner++ has been executed in the logs observing an improvement of the generated process model proportional to the number of activities as time intervals. Figure 11.5 presents results of one particular process, which is mined with different logs (increasing the number of activities expressed as intervals).

In order to aggregate the results in numerical values, we used the F_1 measure, which is described in Sect. 3.1. In particular, *true positives* are the correctly mined dependences; *false positives* are dependences present in the original model but not in the mined one; and *false negatives* are dependences present in mined model but not in the original one.

It is very clear that, even with very small percentages of activities expressed as intervals, there is an important improvement in the mining results.

Application on a Real Scenario

The adaptation of the Heuristics Miner, presented into this section, has been defined starting from some data given by the company Siav S.p.A. We tested our approach against their log. In particular, the original model is the one depicted in Fig. 11.7.

Actually, in this case, all activities are expressed in terms of a set of *sub-activities* (and, in particular, only the start event of each sub-activity is recorded) so, during a pre-processing phase, only the first and last events of each activities were selected, as presented in Fig. 11.8. This phase gives a good approximation of the time intervals,

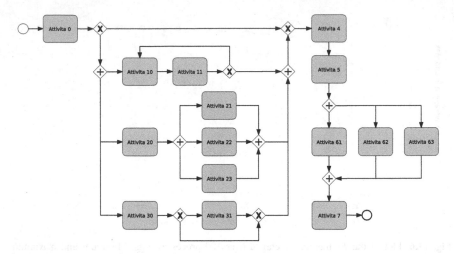

Fig. 11.7 Representation of the process model, by Siav S.p.A., that generated the log used during the test of the algorithm Heuristics Miner++.

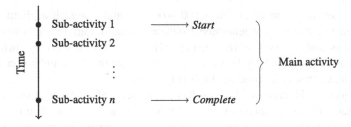

Fig. 11.8 Graphical representation of the preprocessing phase necessary to handle Siav S.p.A. logs.

even if it is not completely correct: the end event represent the start event of the last sub-activity and not the actual end event.

Figure 11.9 shows the result of the mining phase, in which Heuristics Miner++ has been applied. The final model is quite close to the original one and only few edges are not mined correctly. Specifically, the first error is in the dependency between *Attivita0* and *Attivita11*, which is not supposed to appear. The second problem is the missing loop involving *Attivita10* and *Attivita11*. Finally, *Attivita10* should not be connected to *Attivita4* however, by analyzing the graph in details, it is possible to see that this dependency is observed 831 times. Since this value is quite important, we think this is a misbehavior observed in the log.

11.4 Summary

This chapter proposed the description of a generalization of the Heuristics Miner algorithm (i.e., the Heuristics Miner++). This new approach uses the activity expressed

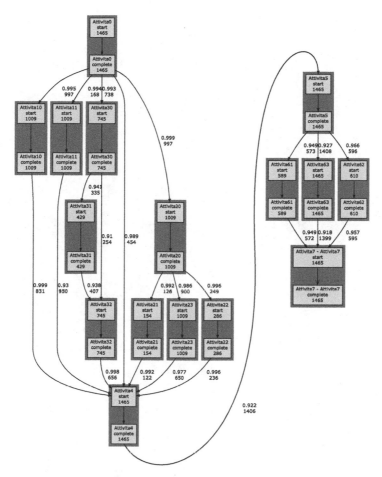

Fig. 11.9 Model mined using Heuristics Miner++ from data generated by model depicted in Fig. 11.7.

as time intervals instead of single events. We introduced this notion into the previous algorithm paying attention to the backward compatibility.

With respect to the problems mentioned in Sect. 1.2, this chapter deals with problem **P-02**: exploiting as much available information as possible, *during* the actual mining phase (see Sect. 8.3). Clearly, there is no final solution for such problem, which depends on the amount of information available in the log. Another approach that will exploit additional information will be reported in Chap. 14.

Chapter 12
Automatic Configuration of Mining Algorithm

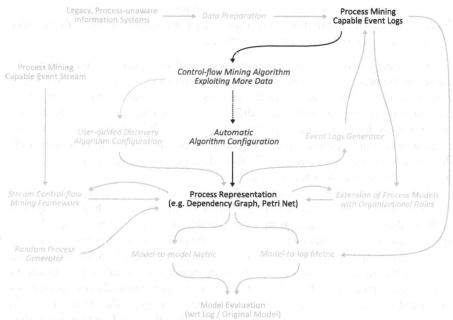

In Sect. 1.2 we mentioned several problems connected to process mining and its application. **P-03** refers to the difficulties in using process mining tools and configuring algorithms. Typical process mining users are non-expert users, therefore it is hard for them to properly configure all the required parameters. This problem has been characterized in Sect. 8.3 as problem *during* the mining phase.

We first describe how certain characteristics of log files raise the need for hand-picking the right parameters of a process discovery algorithm. We then propose a solution to the problem of parameters tuning for the Heuristics Miner++ algorithm. The approach we adopt [23] starts by recognizing that the domain of real-valued parameters can be actually partitioned into a finite number of equivalence classes: so, we suggest to explore the parameters space by a local search strategy driven by

© Springer International Publishing Switzerland 2015   97
A. Burattin: *Process Mining Techniques in Business Environments*, LNBIP 207,
DOI 10.1007/978-3-319-17482-2_12

a Minimum Description Length principle. The proposed result has been then tested on a set of randomly generated process models, obtaining promising results.

12.1 The Problem of Selecting the Right Parameters

When considering real-world industrial scenarios, it is hard to have access to a complete log for a process. In fact, typically the log is partial and/or contains some noise.

We define a log as *partial* if it does not contain a record for all the performed activities; instead, it is *noisy* if either:

1. some activities do not match the "expected" ones;
2. some recorded activities do not match with those actually performed;
3. the order in which activities are recorded may not always coincide with the order in which the activities are actually performed.

While case 1 may be acceptable in the context of workflow discovery, where the names of the performed activities are not set or known *a priori*, cases 2 and 3 may clearly interfere with the mining of the process, leading to an incorrect control-flow reconstruction (that is a control-flow different from the one that the process analyst or the process designer would expect). Because of that, it is important, for a process mining algorithm, to be noise-tolerant. This is especially true for the task of control-flow discovery, where it is more difficult to detect errors because of the initial lack of knowledge on the analyzed process.

A well known example of noise-tolerant control-flow discovery algorithm is Heuristics Miner (and Heuristics Miner++), above-mentioned in Chap. 11. A typical problem that users face, while using these algorithms, is the need to set the values of specific real-valued parameters which control the behavior of the mining, according to the amount and type of noise the user believes is present into the process log. Since the algorithm constructs the model with respect to the number of observations in the log, its parameters basically consist of acceptance thresholds on frequencies of control-flow relevant events, that are observed in the log: if the observed event is frequent enough (i.e., its frequency is above the given threshold for that event) then a specific feature of the control-flow, explaining such event, is introduced. Different settings for the parameters usually lead to different results for the mining, i.e., to different control-flow networks.

While the introduction of these parameters and its tuning is fundamental to allow the mining of noisy logs, the unexperienced user may find difficult to understand the meaning of each parameter and the effect, on the resulting model, of changing the value of one or more parameters from one value to another one. Sometimes, even experienced users find it difficult to decide how to set these parameters.

The approach we propose starts by recognizing that the domain of real-valued parameters can be actually partitioned into a finite number of equivalence classes and we suggest to explore the parameters space by a local search strategy driven by a Minimum Description Length principle. The proposed result is then tested on a set of randomly generated process models, obtaining promising results.

12.2 Parameters of the Heuristics Miner++ Algorithm

The basic measures of Heuristics Miner++ have already been proposed. Here we just list the parameters of the algorithm and, for each of them, a brief description presents the ratio of the specific parameter.

Relative-to-best Threshold. This parameter indicates that we are going to accept the current edge (i.e., to insert the edge into the resulting control-flow network) if the difference between the value of the dependency measure computed for it and the greatest value of the dependency measure computed over all the edges is lower than the value of this parameter.

Positive Observations Threshold. With this parameter we can control the minimum number of times that a dependency relation must be observed between two activities: the relation is considered only when this number is above the parameter value.

Dependency Threshold. This parameter is useful to discard all the relations whose dependency measure is below the value of the parameter.

Length-one Loop Threshold. This parameter indicates that we are going to insert a length-one loop (i.e., a self loop) only if the corresponding measure is above the value of this parameter.

Length-two Loop Threshold. This parameter indicates that we are going to insert a length-two loop only if the corresponding measure is above the value of this parameter.

Long Distance Threshold. With this parameter we can control the minimum value of the long distance measure in order to insert the dependency into the final model.

AND Threshold. This parameter is used to distinguish between AND and XOR splits (if there is more than one connection exiting from an activity): if the AND measure is above (or equal) to the threshold, an AND-split is introduced, otherwise a XOR-split is introduced.

In order to successfully understand the following steps, let's point out an important observation: by definition, all these parameters can have values between -1 and 1 or between 0 and 1. Only the positive observation threshold requires an integer value that expresses the absolute minimum number of observations.

In the first step of Heuristics Miner++, the algorithm extracts all the required information from the process log; then it uses the threshold parameters described above. Specifically, before starting the control-flow model mining, it creates the data structures presented in Table 12.1, where $\mathscr{A}_W$ is the set of activities contained in the log W. All the entries of this data structures are initialized to 0. Then, for each process instance registered into the log, if two activities (a_i, a_{i+1}) are in a direct succession relation, the value of directSuccessionCount$[a_i, a_{i+1}]$ is incremented, while

Table 12.1 Data structures used by Heuristics Miner++ with their sizes. $\mathscr{A}_W$ is the set of activities contained in the log W.

Data structure	Matrix size
directSuccessionCount	$|\mathscr{A}_W|^2$
parallelCount	$|\mathscr{A}_W|^2$
dependencyMeasures	$|\mathscr{A}_W|^2$
L1LdependencyMeasures	$|\mathscr{A}_W|$
L2LdependencyMeasures	$|\mathscr{A}_W|^2$
longRangeDependencyMeasures	$|\mathscr{A}_W|^2$
andMeasures	$|\mathscr{A}_W|^3$

if they are executed in parallel (i.e., the time intervals associated to the two activities overlap) the value of parallelCount$[a_i, a_{i+1}]$ is incremented; moreover, for each activity a, Heuristics Miner++ calculates the length-one loop measure $a \Rightarrow_W a$ and adds its value to L1LdependencyMeasures$[a]$. Then, for each activities pair (a_i, a_j) Heuristics Miner++ calculates the following:

• the dependency measure $a_i \Rightarrow_W a_j$ and then it adds its value to dependency-Measures. It must be noticed that, in order to calculate this metric, the values $|a_i >_W a_j|$ and $|a_i \|_W a_j|$ must be available: these values correspond to the values found in directSuccessionCount$[a_i, a_j]$ and parallelCount$[a_i, a_j]$, respectively;
• the long distance relation measure $a_1 \Rightarrow^l_W a_2$ and adds its value to longRange-DependencyMeasures$[a_1, a_2]$;
• the length 2 loop measure $a_1 \Rightarrow^2_W a_2$ and adds its value to L2Ldependency-Measures.

Finally, for each triple (a_1, a_2, a_3) the procedure calculates the AND/XOR measure $a_1 \Rightarrow_W (a_2 \wedge a_3)$ and adds its value to andMeasures$[a_1, a_2, a_3]$.

When all these values are calculated, Heuristics Miner++ proceeds to the real control-flow construction. These are the main steps: first of all, a node for each activity is inserted; then, an edge (i.e., a dependency relation) between two activities a_i and a_j is inserted if the entry dependencyMeasures$[a_i, a_j]$ satisfies all the constraints imposed by *Relative-to-best Threshold*, *Positive Observations Threshold*, and *Dependency Threshold*.

The algorithm continues iterating through all the activities that have more than one connection exiting from it. It is necessary to disambiguate the split behavior between a XOR and an AND. In these cases (e.g., activity a_i has two exiting connections with activities a_j and a_k), Heuristics Miner++ checks the entry andMeasures$[a_i, a_j, a_k]$ and, if this is above the *AND threshold*, it marks it as an AND-split, otherwise as a XOR-split. If there are more than two activities in the "output-set" of a_i, then all the pairs are checked.

A similar procedure is used to identify length-one loop: Heuristics Miner++ iterates through each activity and checks in the L1LdependencyMeasures vector if the corresponding measure is greater than the *Length-one Loop Threshold*. For the length-two loops the procedure checks, for each activities pairs (a_i, a_j), if `L2LdependencyMeasures[`a_i`,`a_j`]` satisfies the *Length-two Loop Threshold* and, if necessary, adds the loop.

The same process is repeated also for the long distance dependency: for each activity pairs (a_i, a_j), if the value of `longRangeDependencyMeasures` is above the value of the *Long Distance Threshold* parameter, then the dependency between the two activities is added.

Once Heuristics Miner++ has completed all these steps, it can return the final process model. In this case, the final model is expressed as a Heuristics Net (an oriented graph, with information on edges, which can be easily converted into a Petri net).

12.3 Facing the Parameters Setting Problem

As already said, it is not easy for a user (typically process miner users are business process managers, resources managers, or business unit directors) to decide which values to use for the parameters described above: she or he may not be an expert in process mining, and anyway, also an experts in process mining can have an hard time to figure out which setting makes more sense.

The main issue that makes this decision difficult is the fact that almost all parameters take values in real-valued ranges: there is an infinite number of possible choices! Moreover, how can it be possible to select the "right" value for each parameter? Is it preferable to set the parameters in order to generate a control-flow network able to explain all the cases contained in the log (even if the resulting network is very complex and thus hard to understand by a human), or a simpler, and so more readable, model (even if it does not explain all the data)?

Here we assume that the user's desired result of the mining is a "sufficiently" simple control-flow network able to explain as many as possible cases contained in the log. In fact, if the log is noisy, a control-flow network explaining all the cases is necessarily very complex because it has to explain also the noise itself (see [155], for a discussion on this issue).

On the basis of this assumption, we suggest addressing the parameters setting problem by a two step approach:

1. identification of the *candidate hypothesis* that corresponds to the assignments of values to the parameters that induce Heuristics Miner++ to produce different control-flow networks;
2. *exploration* of the hypothesis space to find the "best solution", i.e. generation of the simplest control-flow network able to explain the maximum number of cases.

The aim of step 1 is to identify the set of different process models which can be generated by Heuristics Miner++ by varying the values of the parameters. Among these process models, the aim of step 2 is to select the process model with the best trade-off between complexity of the model description and number of cases that the model is not able to explain. Here, our suggestion is to use the Minimum Description

Length (MDL) [71] approach to formally identify this trade-off. In the next two sections, we describe in detail our definition of these two steps.

12.4 Discretization of the Parameters Values

As discussed in the previous section, by definition, most of Heuristics Miner++ parameters can take an infinite number of values. In practice, only some of them produce a different model as output. In fact, the size of the log used to perform the mining can be assumed to be finite, and thus equations for the various metrics can return only a finite number of different values. These sets, with all the possible values, are obtained by calculating the results of the formulas against all single activities, all pairs, and all triples. Specifically, if we look at the data structures used by Heuristics Miner++, these are populated with all the results just described, so they contain all the possible values of the measures of interest for the given log. Even considering the worst case, i.e. when each activity configuration has a different measure value, the mining algorithm cannot observe more than $|\mathscr{A}_W|^i$ different values for parameters described by an i-dimensional matrix. Since $|\mathscr{A}_W|$ is typically a quite low value, even the worst case does not produce a huge number of possible values. Thus it does not make sense to let the thresholds to assume any real-value in the associated range.

Given a log W, we can sort, in ascending order, all the different values $v_1, \ldots, v_s$, that a given measure can take. Then, all the values in the ranges $[v_i, v_{i+1})$ with $i = 1, \ldots, s$ constitute equivalence classes with respect to the choice of a value for the threshold associated to that measure. In fact, if we pick any value in $[v_i, v_{i+1})$, the output of the mining, i.e. the generated control-flow network, is not going to change. If the parameters were independent, it would be easier to define the set of equivalence classes. In fact, given n independent parameters $p_1, \ldots, p_n$ with domains $D_1, \ldots, D_n$, it is sufficient to compute the set of equivalence classes $\mathscr{E}_{p_i}$ for each parameter p_i, and then obtain the set of equivalence classes over *configurations of the n parameters* as the Cartesian product $\mathscr{E}_{p_1} \times \mathscr{E}_{p_2} \times \cdots \times \mathscr{E}_{p_n}$. This means that we can uniquely enumerate process models by tuples $(d_{1,i_1}, \ldots, d_{n,i_n})$, where $d_{j,i_j} \in D_j$, $j = 1, \ldots, n$.

Unfortunately, by definition, Heuristics Miner++ parameters are not independent. This is clearly exemplified by considering only the two parameters *Positive Observations Threshold* and *Dependency Threshold*. If the first one is set to a value that does not allow a particular dependency relation to appear in the final model (because it does not occur frequently enough in the log), then, there is no value for the dependency threshold, involving the excluded dependency relation, that will modify the final model. The lack of independence entails that the mining procedure may generate exactly the same control-flow network starting by different settings for the parameters. This means that it is impossible to uniquely enumerate all the different process models by defining the equivalence classes over the parameters values, as discussed above under the independence assumption. So, since there is not a bijective function between process models and tuples of discretized parameters, it is not possible to efficiently search the "best" model by searching among the discretized space of parameters. However, discovering all the dependences among the parameters and then defining a restricted tuple space where there is a one to one correspondence between tuples and process models would be difficult and expensive. Therefore, we decided to adopt the independence assumption to generate the tuple space, while using high level knowledge about the dependences among parameters to factorize the tuple space, in order to perform an approximate search.

12.5 Exploration of the Hypothesis Space

We have just described a possible way to discover distinct process models produced by a set with all the values for each parameter. As we have discussed before, each process model mined from a particular parameters configuration constitutes, for us, a hypothesis (i.e. a potential candidate to be the final process model). We are, now, in this situation: *(a)* it is possible to build a set with all possible parameters values; *(b)* each parameters configuration produces a process model hypothesis. Starting from these two elements, we can realize that we have all the information required for the construction of the hypothesis space: if we enumerate all the tuples of possible parameters configurations (and this is possible, since these sets are finite), we can build the set of all possible hypotheses, which is the hypothesis space. The second step, described in our approach, requires the exploration of this space, in order to find the "best" hypothesis.

In order to complete the definition of our search strategy, it is necessary to finally give a formal definition of our measure of "goodness" for a process model. To this aim, we adopt the Minimum Description Length (MDL) principle [71]. MDL is a widely known approach, based on the Occam's Razor: *"choose a model that trades-off goodness-of fit on the observed data with 'complexity' or 'richness' of the model"*. Let's take as an example the problem of communicating through a very expensive channel: we can build a compression algorithm whereby the most frequent words are represented in the shortest way, while the less frequent have a longer representation. Now, as first thing to do, we have to transmit the algorithm, then we can use it to send our encoded messages. We have to pay attention in not building a too complex (meaning that can handle many cases) algorithm: its transmission may neutralize the benefits of its use, in terms of total amount of data to be transmitted. Consider now the set H of all possible algorithms that can be built and, given $h \in H$, let $L(h)$ be its description size and $L(D \mid h)$ will be the size of the message D after its compression using h. The MDL principle tells us to choose the "best" hypothesis h_{MDL} as:

$$h_{MDL} = \arg\min_{h \in H} L(h) + L(D \mid h).$$

In [31], Calders *et al.* present a detailed approach to compute Minimum Description Length for process mining. In this case, the model is always assumed to be a

Petri net. Specifically, the proposed metric shows two different encodings, for the model and for the log:

- $L(h)$ is the encoding of the model h (a Petri net), and lies in a sequence of all the elements of the net (i.e. places and transitions). For each place, moreover, the sets of incoming and outgoing transitions are recorded too. The result is a sequence structured as: $\langle$transitions, places (with connections)$\rangle$.
- $L(D \mid h)$ represents the encoding of the log D and is a bit more complex. Specifically, the basic idea is to replay the entire log on the model and, every time an error occurs (i.e. the event of the log cannot be replayed by the model), a "punishment" is assigned. The approach punishes also the case in which there are too many transitions enabled at the same time (in order to avoid models similar to the "flower model", see Fig. 6.1(b)).

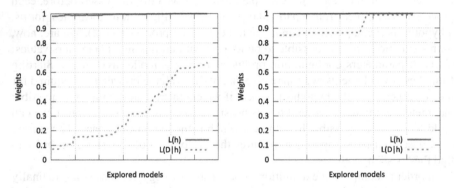

Fig. 12.1 Unbalancing between different weights of $L(h)$ and $L(D \mid h)$ according to the MDL principle described in [31]. The left hand side figure shows an important discrepancy, the right hand one does not.

The same work proposes to weight the two encodings according to a convex combination, so to let the final user decide how to balance the two weights. We used this approach to guide the search of the best hypothesis. However, several problems limit the use of such approach. The most important ones are:

- the reply of the traces is very expensive from a computational point of view. The approach resulted absolutely unfeasible in industrial scenarios, with "real data". For example, after performing several optimizations and executing a simple model in a controlled environment, the procedure required up to 20 h for running[1];
- the codomain of the values of the two measures ($L(h)$ and $L(D \mid h)$) is not actually bounded (even after the normalization proposed on the plugin implementation[2]). Moreover, in our examples, we observed that the values of $L(h)$ and $L(D \mid h)$

---

[1] These experiments are also reported on the M.Sc. thesis by D. Turato: "*Configurazione automatica di Heuristics Miner++ tramite il principio MDL*", at the University of Padua, Italy.

[2] See http://www.processmining.org/online/mdl for more information.

are very unbalanced, therefore their averaging is not really producing expected effects. An example of this problem is reported in Fig. 12.1[3].

Because of these problems, we "relaxed" the measures of the model and of the data, so to have lighter versions of them, capable of capturing the concepts we need.

12.6 Improved Exploration of the Hypothesis Space

The parameters discretization process does not produce a large number of possible values but, since the hypothesis space is given by the combination of all the parameters values, this can become quite large, and finding the best hypothesis easily turns into a quite complex search problem: an exhaustive search of the hypothesis space (that will lead to the optimal solution) is not feasible. So we decided to factorize the search space by exploiting high level knowledge about independent relations (both total and conditional) among parameters, and to explore the so factorized space by a local search strategy. We are now going to describe the factorization of the search space.

12.6.1 Factorization of the Search Space

Heuristics Miner++ parameters are not independent. Their dependencies can be characterized by listing the main operations performed by the mining algorithm, and the corresponding parameters:

1. calculation of the length-one loops and check *Length-one Loop Threshold* and *Positive Observations Threshold*;
2. calculation of the length-two loops and check *Length-two Loop Threshold* and *Positive Observations Threshold*;
3. calculation of the dependency measure and check *Relative-to-best Threshold*, *Positive Observations Threshold* and *Dependency Threshold*;
4. calculation of AND measure and check *AND Threshold*;
5. calculation of long distance measure and check *Long Distance Threshold*.

When more than one parameter is considered within the same operation, all the corresponding checks have to be considered as in conjunction relation, meaning that all constraints must be satisfied. The most frequent parameter that is verified is the *Positive Observations Threshold*, occurring in three steps; under these conditions, if, as an example, the dependency relation under consideration does not reach a sufficient number of observations in the log, then the check of parameters *Relative-to-best Threshold, Dependency Threshold, Length-one Loop Threshold* and *Length-two Loop Threshold* can be skipped because the whole check (the 'and' with all other parameters) will not pass, regardless of the success of the single checks involving the *Relative-to-best Threshold*, the *Dependency Threshold*, the *Length-one Loop Threshold* and the *Length-two Loop Threshold*.

---

[3] See footnote 1.

Besides that, there are some other intrinsic rules on the design of the algorithm: the first says that if an activity is detected as part of a length-one loop, then it can't be in a length-two loop and *vice versa* (so, checks in step 1 and step 2 are in mutual exclusion); another tells that if an activity has less than two exiting edges, then it is impossible to have an AND or XOR split (and, in this case, step 4 does not need to be performed).

In order to simplify the analysis of the possible mined networks, we think it is useful to distinguish two types of networks, based on the structural elements they are composed of:

- *Simple networks*, which include process models with no loops and no long distance dependencies;
- *Complete networks*, which include simple networks extended with at least one loop and/or one long distance dependency.

For the creation of the first type of networks, only steps 3 and 4 (on the list at the beginning of this section) are involved: so only *Relative-to-best Threshold*, *Positive Observations Threshold*, *Dependency Threshold* and *AND Threshold* have an important role in the creation of this class of networks. Complete networks are obtained by adding, to a simple network, one or more loops, by using steps 1 and 2, and/or one or more long distance dependencies via step 5. It can be observed that, once the value for *Positive Observations Threshold* is fixed, steps 1, 2, and 5, are in practice controlled independently by *Length-one Loop Threshold*, *Length-two Loop Threshold*, and *Long Distance Threshold* respectively.

12.6.2 Searching for the Best Hypothesis

At this point, the new objective is the definition of the process for the identification of the "best" model (actually, we have to find the best parameters configuration). There are two issues here: the first is the definition of some criterion to determine what means "best model". Secondly, we will have to face the problem of the hypothesis space that is too big to be exhaustively explored. We are going to start from the latter problem, assuming to have a criterion to quantify the goodness of a process model.

For what concerns the big dimension of the search space, we start the search within the class of simple networks and, once the system finds the "best" local model, it tries to extend it into the complete network space. With this division of the work, the system reduces the dimensionality of the search spaces.

From an implementing point of view, in the first phase, the system has to inspect the joint space composed only of *Relative-to-best Threshold*, *Positive Observations Threshold*, *Dependency Threshold* and *AND Threshold* (the parameters involved in "simple networks") and, when it finds a (potentially only local) optimal solution, it can try to extend it introducing loops and long dependency. In Fig. 12.2 we propose a graphical representation of the main phases of the exploration strategy. Of course, this search strategy is not complete for two reasons: *(i)* local search is, by definition, not complete; and *(ii)* the "best" process model may be obtained by extending with loops and/or long dependencies a sub-optimal simple network.

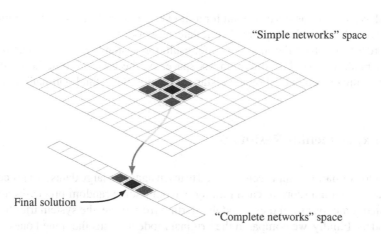

Fig. 12.2 Graphical representation of the searching procedure: the system looks for the best solution on the *simple network* class. When a (local) optimal solution is found, the system tries to improve it by moving into the *complete network* space. Red boxes indicate the neighborhood of the current node which is drawn in black (Color figure online).

Concerning the actual length measures, we studied, as model complexity $L(h)$, the number of edges in the network. This is an easily computable measure, although it may underestimate the complexity of the network, because it disregards the different constructs that compose the network. Anyway, this is a good way to characterize the description length of the process model.

As $L(D \mid h)$ measure, we use the fitness measure introduced in [152] and, in particular, we opted for the *continuous semantics* one. Differently from the *stop semantics*, the one chosen does not stop at the first error, but continues until it reaches the end of the model. This choice is consistent with our objective to evaluate the whole process model. This measure is expressed as:

$$f_{M,W} = 0.4 \cdot \frac{parsedActs(M, W)}{|\mathscr{A}_W|} + 0.6 \cdot \frac{parsedTraces(M, W)}{logTraces(W)}$$

where M is the current model and W, as usual, is the log to "validate"; $|\mathscr{A}_W|$ is the number of activities in the log and $logTraces(W)$ is the number of traces in W; $parsedActs(M, W)$ gives the sum of all parsed activities for all traces in W and $parsedTraces(M, W)$ returns the number of traces in W completely parsed by the model M (when the final marking involves only the last activity).

The search algorithm starts from a random point in the "simple network" space and, exploiting a hill-climbing approach [134], evaluates all the neighbor simple networks obtained by moving the current value of one of the parameters up or down of a position within the discretized space of possible values. If a neighbor network (with a better MDL value) exists, then that network becomes the current one and the search is resumed until no better network is discovered. The "optimal" simple network is then used as starting point for a similar search in the remaining parameters space, so to discover the "optimal" complete network, if any.

In order to improve the quality of the result, the system restarts the search from another random point in the hypothesis space. At the end, only the best solution among all the ones obtained by the restarts is proposed as "final optimal" solution.

12.7 Experimental Results

In order to evaluate our approach, we tried to test it against a large dataset of processes. In order to assign a score to each mining, we built some random processes and we generated some logs from these models; starting from these, the system tries to mine the models. Finally, we compared the original models versus the mined ones.

12.7.1 Experimental Setup

The set of processes to test is composed of 125 process models. These processes were created using the approach presented in Chap. 16.

The generation of the random processes is based on some basic "process patterns", like the AND-split/join, XOR-split/join, the sequence of two activities, and so on. In Fig. 12.3 some statistical features of the dataset are shown. For each of the 125 process models, two logs were generated: one with 250 traces and one with 500 traces. In these logs, the 75 % of the activities are expressed as time intervals (the other ones are instantaneous) and 5 % of the traces are noise. In this context, "noise" is considered either a swap between two activities or removal of an activity.

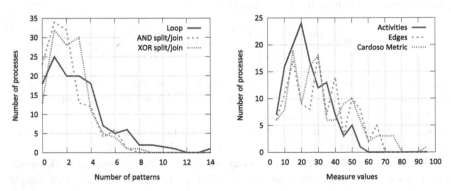

Fig. 12.3 Features of the processes dataset. The left hand side plot, reports the number of processes with a particular number of patterns (AND/XOR splits/joins and loops). The plot in the right hand side contains the same distribution versus the number of edges, the number of activities and the Cardoso metric [32] (all these are grouped using bins of size 5).

We tried the same procedure under various configurations: using 5, 10, 25 and 50 restarts. In the implemented experiments, we run the algorithm allowing 0, 1, 5 and 10 lateral step, in case of local minimum (in order to avoid problems in case of very small plateau).

The distance of the mined process from the correct one is evaluated with the F_1 measure (see Sect. 3.2).

12.7.2 Results

The number of improvement steps performed by the algorithm is reported in Fig. 12.4. As shown in the figure, if the algorithm is run with no lateral steps, then it stops early. Instead, if lateral steps are allowed, the algorithm seems to be able, at least in some cases, to get out of plateaus. In our case, even 1 step shows a good improvement in the search. The lower number of improvement steps (plot on the right hand side), in the case of 500 traces, is due to the fact that, with more cases, it is easier to reach an optional solution.

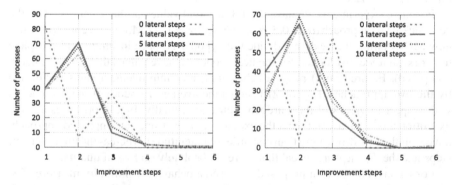

Fig. 12.4 Number of processes whose best hypothesis is obtained with the plotted number of steps, under the two conditions of the mining (with 0, 1, 5 and 10 lateral steps). The left hand side plot refers to the processes mined with 250 traces while the right hand side refers to the mining using 500 traces.

The quality of the search mining result, as measured by the F_1 measure, is shown in Fig. 12.5. Results for 250 traces are reported in the left hand side plot, while results for 500 traces are shown in the right hand side plot. It is noticeable that the average F_1 is higher in the 500-traces case. This phenomenon is easily explainable since, with a larger dataset, the miner is able to extract a more reliable model.

Several tests have been performed considering also the MDL approach described in [31] (presented in Sect. 12.5)[4]. However, due to the time required for the processing

---

[4]See footnote 1.

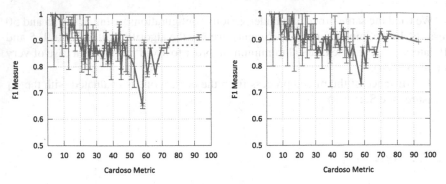

Fig. 12.5 "Goodness" of the mined networks, as measured by the F_1 measure, versus the size of the process (in terms of Cardoso metric). The left hand size plot refers to the mining with 250 traces, while the right hand side plot refers to the mining with 500 traces. Dotted horizontal lines indicate the average F_1 value.

of the entire procedure, we considered only a fraction of our dataset: 93 process models (the simplest ones), logs with only 250 traces and with no noise. Results are reported in Fig. 12.6.

For these experiments we have tried 3 different values of the α parameter of the "classic" MDL approach ($\alpha = 0.3, \alpha = 0.5$, and $\alpha = 0.7$). Moreover, concerning our new MDL definition, we do not divide the hypothesis space in simple and complete networks, but we just looked for the best model (to have values that can be reliably compared). Figure 12.6(a) proposes the number of improvement steps performed by the two approaches; Fig. 12.6(b) shows the average F_1 score of the approaches with respect to values of the Cardoso metric and, finally, Fig. 12.6(b) presents the execution times. Please note that the execution times, using our improved approach, have significantly drop (more than one order of magnitude), whereas the improvement steps and the F_1 measure reveal that there is absolutely no loss of quality.

For the last comparison proposed, we used a behavioral similarity measure. The idea underpinning this measure, which will be presented in details in Subsect. 15.1.3, is to compare all the possible dependencies that the two processes allow and all the dependencies that are not. Therefore, the comparison is performed according to the actual behaviors of the two processes, independently of their edges. Such approach, differently from the F_1, is also able to discriminate AND and XOR connections. Figure 12.7 shows the similarity values of all the models. In this case (as in the previous ones), we do not divided the set of models in simple and complete networks. As it could be seen, our improved approach is not penalized in any way, with respect to the well founded MDL executions. Instead, for several processes it seems to be able to obtain even better models.

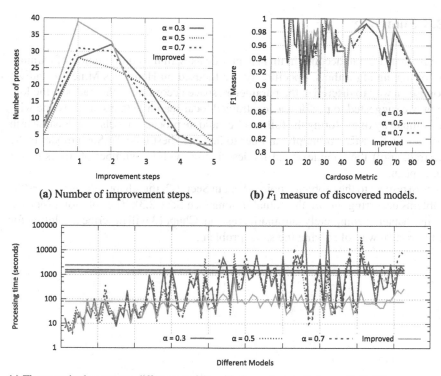

(a) Number of improvement steps.   (b) F_1 measure of discovered models.

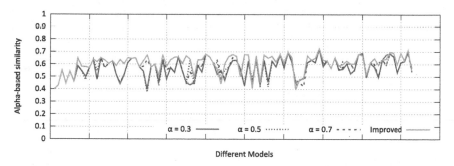

(c) Time required to process different models, with the various techniques. Dotted lines represent the averages of each approach. Logarithmic scale is used.

Fig. 12.6 Comparison of results considering the classical MDL measures and the improved ones. These results refer to runs with 10 lateral steps and 10 random restarts.

Fig. 12.7 Performance comparison in terms of Alpha-based metric. These results refer to runs with 10 lateral steps and 10 random restarts.

12.8 Summary

The issue taken into account in this chapter deals with the configuration of parameters of mining algorithms. Specifically, we focused on Heuristics Miner++ and we proposed a way to discretize the parameters space according to the traces in the log. Then, we suggested to perform a constrained local search in that space to cope with the complexity of exploring the full set of candidate process models. The local search is driven by the MDL principle in order to find the "best model". Such model is the one trading-off the complexity of its description, with the number of traces that it can explain.

With respect to the problems pointed out in Sect. 1.2, this chapter deals with **P-03**: problems occurring *during* the actual mining (see Sect. 8.3). The solution proposed in this chapter is completely automatic; instead, Chap. 13 will describe a solution for an interactive way of solving the same problem.

Chapter 13
User-Guided Discovery of Process Models

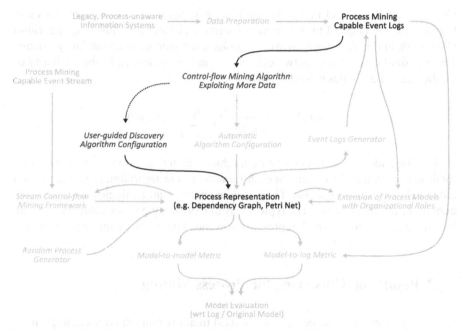

As mentioned in Sect. 1.2, problem **P-03** deals with difficulties in using process mining tools and configuring algorithms (see Sect. 8.3). In Chap. 12 we presented a completely automatic approach to solve **P-03**. This chapter, instead, presents a more flexible solution, which requires the interaction of a user.

The approach that will be described shifts the problem from choosing the best parameters configuration to selecting the model that better describe the actual process performed. This is the main reason why such approach can also be called "*parameter configuration via result exploration*". The idea can be split in the following steps:

1. the system receives a log as input;
2. the space of the parameters can be discretized in order to consider only the meaningful values (from an infinite space to a finite one);

© Springer International Publishing Switzerland 2015
A. Burattin: *Process Mining Techniques in Business Environments*, LNBIP 207,
DOI 10.1007/978-3-319-17482-2_13

3. all the distinct models that could be generated starting from the parameters are eventually build, so to have an exhaustive list of all the models that can be inferred starting from the log;
4. all the generated processes are clustered;
5. the hierarchy of clusters is "explored" by the user by drilling down on the direction that he/she thinks being the most promising one.

A practical example of the above-mentioned approach is presented in the following sections.

13.1 Clustering for Process Mining

Once a model-to-model metric is available, it is possible to cluster processes in hierarchies. We decided to use an agglomerative hierarchical clustering algorithm [103] with, in this first stage, an average linkage (or average inter-similarity): in this case the similarity s between two clusters, c_1 and c_2, is defined as the similarity of all the pairs of activities belonging to the two clusters:

$$s(c_1, c_2) = \frac{1}{|c_1||c_2|} \sum_{p_i \in c_1} \sum_{p_j \in c_2} d(p_i, p_j)$$

The basic idea of agglomerative hierarchical clustering is to start with each element in a singleton cluster and, at each iteration of the algorithm, to merge the two closest cluster into a single one. The procedure iterates until a single cluster remains, containing all the elements. The typical way of representing a hierarchical clustering is using a dendrogram, which represents how the elements are combined together.

13.2 Results on Clustering for Process Mining

Clustering of business processes can be used to allow non-expert users to perform process mining (as control-flow discovery). A proof of concept procedure has been implemented.

The approach has been tested on a process log with 100 cases and 46 event classes, equally distributed among each case, with 2 event types. The complete set of possible business process is made of 783 models that were generated starting from the possible configurations of the algorithm Heuristics Miner++.

A complete representation of the clusters generated from such dataset has not been created because of problems in exporting the image, however, a representation of a subset of them (350 process models) is proposed in Fig. 13.1. This is a dendrogram representation of the hierarchy that comes out of the distance function presented in

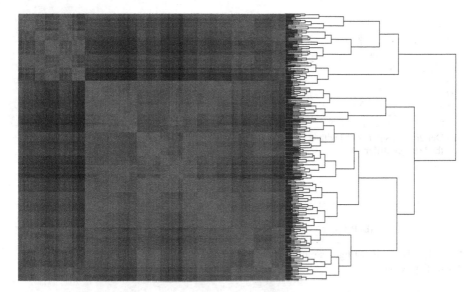

Fig. 13.1 Distance matrix of 350 process models, generated as different configuration of the Heuristics Miner++ parameters. The brighter an area is, the higher is the similarity between the two processes (e.g., the diagonal). The dendrogram generated starting from the distance matrix is proposed too.

previous sections. The distance matrix, with distances per each pair of models, is presented as well.

Concerning the approach "parameter configuration via result exploration", the idea is to start from the "root" of the dendrogram and "navigate" it until a leaf is reached. Since a dendrogram is a binary tree, every cluster is made of two sub-clusters that are represented by their corresponding medoids. These two process models (i.e., the medoids) are proposed, at each step, to the user, who can decide which is the best "direction" to follow. In the first steps, the user will be asked to select between models that are very different each other. As long as the user takes decisions, the processes to compare will be closer each other, so the user decision can be based on very different and detailed aspects.

Figure 13.2 reports the dendrogram with $\alpha \in \{0, 0.5, 1\}$. Hierarchical clustering has been performed on 10 randomly generated business processes. The result is presented in the figure. In the lower part of the same figure examples of two processes considered "distant" are also reported.

13.3 Implementation

All the techniques described in this book have been implemented in ProM. Both version 5 [173] and 6 [179] have been used.

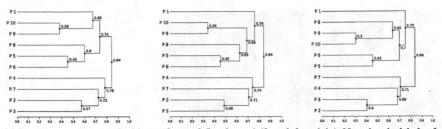

(a) Dendrograms generated with $\alpha = 0$, $\alpha = 0.5$ and $\alpha = 1$ (from left to right). Note that the labels of the leafs are different.

(b) P 4.   (c) P 10.

Fig. 13.2 The topmost figures represent three dendrograms. The two Petri nets are examples of "distant" processes.

ProM[1] is a framework which can be extended and used through a plugin architecture. Several plugins are available, implementing a series of process mining algorithms. The main advantage in using ProM consists in having all the basic operations (e.g. log and models input/output and graphic visualizers) available in a single and open-source framework. Starting from ProM 6, the default input format for log files is XES [75][2] (Listing 17.1 proposes a fragment of XES code), but MXML [73] is still supported as well. ProM is platform independent and it is written in Java. Currently, ProM is maintained by the Process Mining Group[3] of the Eindhoven Technical University[4].

The Heuristics Miner++ algorithm, described in Chap. 11, has been implemented in ProM 5. Figure 13.3 proposes a couple of screenshots of the implementation of Heuristics Miner++. The same figure proposes a visualization of the parameter discretization. The implementation can be downloaded from http://www.processmining.it/sw/hmpp.

The automatic approach, described in Chap. 12, has been implemented in ProM 5 but only as a command line application, since it is supposed to periodically run in an autonomous manner.

Apart from this mining plugins, a "time filter plugin" has been implemented in ProM 6, as presented in Fig. 13.4. The basic idea is to present a log as a dotted chart [142] (not dots for events, but lines for traces). It is possible to sort the log according

---

[1] See http://www.promtools.org/ for more information.

[2] The IEEE Task Force on Process Mining Meeting, at the BPM 2012 meeting, decided to start the procedure to let XES (http://www.xes-standard.org/) become an IEEE standard (http://standards.ieee.org/).

[3] http://www.processmining.org/.

[4] http://www.tue.nl/.

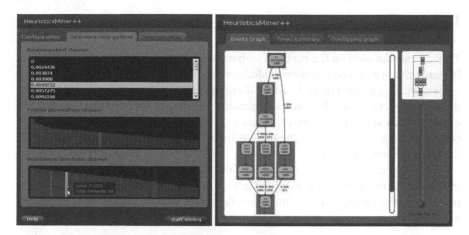

Fig. 13.3 The figure on the left hand side contains the configuration panel of Heuristics Miner++, where parameters are discretized. The screenshot on the right hand side shows the result of the mining.

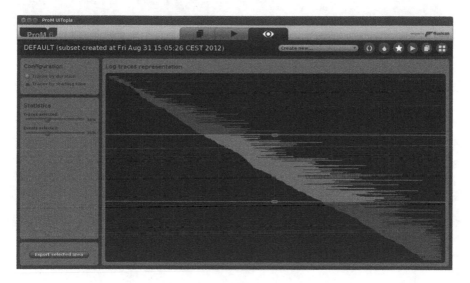

Fig. 13.4 Time filter plugin, with traces sorted by starting point and with a fraction of the log selected.

to the traces duration or according to the trace starting time. Using two "sliders", the analyst can select a subset of the current log. Moreover, the plugin gives information on the percentage of traces and events selected: this allows you, for example, to select only the top 10 % longest traces. Finally, the selected traces can be "exported" in order to be analyzed independently from the rest of the log, to get more insights on particular cases (e.g. outliers).

13.4 Summary

This chapter focused on the problem of parameters setting for Heuristics Miner++. In particular, we proposed a user-guided approach.

Specifically, given a discretization of the parameter values, the user-guided configuration is actually an alternative approach to explore the space of models: the user explores such space of processes through the medoids of the clusters resulting as output of the generation of all the models (obtained performing the mining with all the different configurations).

With respect to the problems pointed out in Sect. 1.2, this chapter deals with **P-03**: problems occurring *during* the actual mining (see Sect. 8.3). The solution proposed in this chapter requires the interaction of the user; instead, Chap. 12 reported a completely automatic solution to tackle the same problem.

Chapter 14
Extensions of Business Processes with Organizational Roles

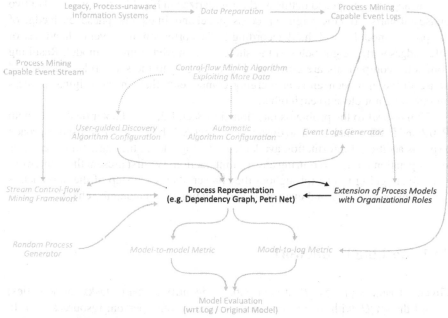

Chapter 5 presents the three basic types of process mining, which are also described in Fig. 5.1. Moreover, as stated in Sect. 5.2, several *perspectives* might be involved in process mining. Specifically, it is possible to concentrate on:

- the *control-flow*, which is a (possibly graphical) representation of the business process model (i.e., the ordering of activities);
- the *organizational perspective*, which focuses on the interactions among activities originators;
- focusing on *cases* (single process instances) may help identifying peculiarities based on specific characteristics (for example, which case conditions lead to a particular path of the process model);

A. Burattin: *Process Mining Techniques in Business Environments*, LNBIP 207,
DOI 10.1007/978-3-319-17482-2_14

- the *time perspective* is extremely useful to measure and monitor the process, for example to find bottlenecks or predict the remaining time of a case.

In this chapter, we concentrate on the *extension* of a process from the *organizational perspective*. Specifically, we present an approach [29] which, given a process model and a log in input, tries to partition the set of activities of the process into "swimlanes". This partitioning is performed by grouping originators in roles and associating activities with the corresponding role.

The approach proposed in this chapter is based on the identification of roles and this is, in turn, based on the observation of the distribution of originators over activities and roles. This division is extremely important and gives new detailed insights on the process model (which can be extracted using discovery techniques). For example, it is possible to compare the actual roles distribution with the mined ones or to analyze the proposed roles in order to improve the current organization.

The approach proposed in this work, summarized in Fig. 14.1, is composed of two phases: it starts from the original process model and, in the first phase, each edge of the process model is weighted according to the corresponding level of handover of role. Edges with weight below a threshold are removed from the model. Resulting connected components are considered as belonging to the same role. The second phase of the approach aims at merging components that, in the original process model, were not close to each other.

With respect to the problems mentioned in Sect. 1.2, this chapter deals with both **P-02** and **P-03**, which are also discussed in Sect. 8.3. In fact, the proposed approach exploits additional information available in the log and, at the same time, performs such operation without requiring any user interaction. We first present the framework that we are going to use throughout the chapter, then each step of the approach is separately presented. Final evaluation and a summary conclude the chapter.

14.1 Working Framework

Given a business process P, it is possible to identify its set of tasks (or activities) A and the set U with all the involved originators (e.g. person, resources, ...). In this context, the complete set of observable events, generated by P, is defined as $E = A \times U$.

A process can generate a log $L = \{e_1, \ldots, e_n\}$, which is defined as a set of traces. Each element of the log identifies a case (i.e. a process instance) of the observed model. A trace $\mathbf{e} = \langle e_1, \ldots, e_m \rangle$ is a sequence of events, where $e_j \in E$ represents the jth event of the sequence. With $e_i \in \mathbf{e}$ we indicate that event e_i is contained in the sequence $\mathbf{e}$.

Given a process model P, let $\mathcal{D}(P)$ be the set of direct dependencies (i.e. directed connections) of the process model. For the sake of simplicity, whenever there is no ambiguity on the process P, we assume $\mathcal{D}$ as a synonym of $\mathcal{D}(P)$. For example, the set $\mathcal{D}$ of the process model depicted in Fig. 14.1(a) is: $\mathcal{D} = \{A \to B,$

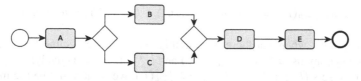

(a) Original process model.

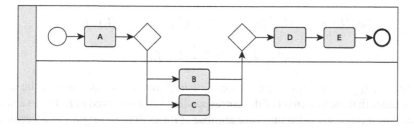

(b) Expected result, with activities partitioned in roles.

Fig. 14.1 Input and expected output of the approach presented in this chapter.

$A \rightarrow C, B \rightarrow D, C \rightarrow D, D \rightarrow E$}. We assume to have the possibility to "replay" activities of traces on the process model (e.g. [150] proposes an approach for replay).

Given an event $e \in E$, such that $e = (a, u)$, let's define the typical projection operators $\pi_A(e) = a$ and $\pi_U(e) = u$. Moreover, let us define the operator $U_a(L)$ as:

$$U_a(L) = \{\pi_U(e) \mid \exists_{e \in L} \, e \in \mathbf{e} \land \pi_A(e) = a\}.$$

Given a dependency $a \rightarrow b \in \mathcal{D}$, it is possible to define the set of couples of originators $U_{a \rightarrow b}(L)$:

$$U_{a \rightarrow b}(L) = \{(\pi_U(e_i), \pi_U(e_j)) \mid \text{the replay algorithm identifies a}$$
$$\text{dependency of } a \rightarrow b \text{ mapped to } e_i \text{ and } e_j\}.$$

This operator returns the set of couples of originators, in the log L, that performs the dependency $a \rightarrow b$.

Similar operators are $U^a_{a \rightarrow b}(L)$ and $U^b_{a \rightarrow b}(L)$. They can be used to obtain originators of activity a or b, when they are involved in the dependency $a \rightarrow b$:

$$U^a_{a \rightarrow b}(L) = \{u_i \mid (u_i, u_j) \in U_{a \rightarrow b}(L)\},$$

$$U^b_{a \rightarrow b}(L) = \{u_j \mid (u_i, u_j) \in U_{a \rightarrow b}(L)\}.$$

On all these sets, it is possible to apply the classical Relational Algebra operators [53]. For example, with the selection operator it is possible to define:

$$\sigma_=(U_{a \to b}(L)) = \{(u_i, u_j) \mid (u_i, u_j) \in U_{a \to b}(L) \wedge u_i = u_j\}.$$

For simplicity, whenever there is no ambiguity on L, we assume $U_a, U_{a \to b}$ and $U^a_{a \to b}$ as a synonyms of $U_a(L)$, $U_{a \to b}(L)$ and $U^a_{a \to b}(L)$, respectively.

Given the sets $U_a(L)$, $U_{a \to b}(L)$, and $U^a_{a \to b}(L)$, we want to define the multisets [144] $\mathcal{U}_a(L)$, $\mathcal{U}_{a \to b}(L)$, and $\mathcal{U}^a_{a \to b}(L)$ which take into account the frequency of the originators in L:

$$\mathcal{U}_a(L) = \langle U_a(L), f_{U_a} \rangle \qquad \mathcal{U}_{a \to b}(L) = \langle U_{a \to b}(L), f_{U_{a \to b}} \rangle$$

$$\mathcal{U}^a_{a \to b}(L) = \langle U^a_{a \to b}(L), f_{U^a_{a \to b}} \rangle$$

where f_{U_a}, $f_{U_{a \to b}}$, and $f_{U^a_{a \to b}}$ are the *multiplicity functions*, which indicate the number of times that each element of the corresponding set is observed in L. For example, given $u \in \mathcal{U}_a(L)$, $f_{U_a}(u)$ returns the number of times that the originator u performs activity a in L. In this work, the cardinality of a multiset $\mathcal{M} = \langle M, f_M \rangle$ is defined as the sum of the values of the multiplicity function, for the elements of the multiset:

$$|\mathcal{M}| = \sum_{m \in M} f_M(m).$$

The intersection of two multisets $\mathcal{M}_1 = \langle M_1, f_{M_1} \rangle$ and $\mathcal{M}_2 = \langle M_2, f_{M_2} \rangle$ is defined as the intersection of the two sets M_1 and M_2 and the multiplicity function is defined as the minimum between the multiplicity values:

$$\mathcal{M}_1 \cap \mathcal{M}_2 = \langle M_1 \cap M_2, \min\{f_{M_1}(x), f_{M_2}(x)\} \rangle.$$

In this context, we will also consider the sum of multisets. Given $\mathcal{M}_1 = \langle M_1, f_{M_1} \rangle$ and $\mathcal{M}_2 = \langle M_2, f_{M_2} \rangle$, the sum is defined as:

$$\mathcal{M}_1 \uplus \mathcal{M}_2 = \langle M_1 \cup M_2, f_{M_1}(x) + f_{M_2}(x) \rangle.$$

For the sake of simplicity, we will omit L whenever there is no ambiguity (e.g., $\mathcal{U}_a$ instead of $\mathcal{U}_a(L)$). Moreover, the notation $\mathcal{M} = \{a^x, b^y\}$ identifies the multiset where a has multiplicity x and b has multiplicity y.

The selection operator σ_θ can be used also on multisets. For example, $\sigma_=(\mathcal{U}_{a \to b}) = \langle \sigma_=(U_{a \to b}), f_{U_{a \to b}} \rangle$ (where the multiplicity function is defined only on elements of the set $\sigma_=(U_{a \to b})$).

The problem we will try to solve is to find a partition [18] $\mathbf{R} \subset \mathcal{P}(A)$[1] of the set of activities A, given a log L and the original process P, such that:

- $\bigcup \mathbf{R} = A$, i.e., the partitioning $\mathbf{R}$ *covers* the entire set of tasks A; and
- for each $X, Y \in \mathbf{R}$, such that $X \neq Y$, $X \cap Y = \emptyset$, i.e., all the partitions are pairwise disjoint.

---

[1] $\mathcal{P}(A)$ identifies the powerset of A.

From the business point of view, we are requiring that each activity needs to belong to exactly one role. The partition **R** identifies the set of roles of the process. In this context, the term "partition of activities" and "role" are used as synonyms.

Let $|L|$ be the size of the log, i.e., the number of traces it contains. Given a log L and an originator $u \in U$, we define $|L|^u$ as:

$$|L|^u = \sum_{e \in L} \sum_{i=1}^{|e|} |\{e_i \mid \pi_U(e_i) = u\}|.$$

In other words, $|L|^u$ returns the number of times that originator u executes activities in L. A similar measure, which also takes into account the role, is $|L|^u_R$, where u is an originator and R is a role (i.e. a set of activities):

$$|L|^u_R = \sum_{e \in L} \sum_{i=1}^{|e|} |\{e_i \mid \pi_A(e_i) \in R \wedge \pi_U(e_i) = u\}|.$$

Finally, given a log L and a partition R, it is possible to define the multiset of originators involved in the role as:

$$\mathcal{U}_R(L) = \biguplus_{a \in R} \mathcal{U}_a(L).$$

As presented in Sect. 14.5, approaches for the identification of the handover of work between originators already exist; however, this work proposes an approach to point out *handover of roles* and therefore the identification of roles themselves. This operation is based on who are the activity originators. Specifically, we assume that, under ideal scenarios, there is a clear distinction among originators performing activities belonging to different roles. However, it is really difficult to observe such clear distinction in business environments (i.e., originators are involved in several roles) and thus we need to resort to a metric to measure the degree of handover between roles. This, and how to define a role, are the topics covered by the next section.

14.2 Rules for Handover of Roles

As stated in the previous section, the identification of business roles, as presented in this work, assumes that an activity is not allowed to belong to different roles at the same time. Let us recap: given a process P and the dependency $a \to b \in \mathcal{D}(P)$, $\mathcal{U}^a_{a \to b}(L)$ is the multiset of originators (with frequencies) that perform the activity a (as part of the dependency $a \to b$) in the log L; and $\mathcal{U}_{a \to b}(L)$ identifies the set of couples of originators (with frequencies) performing a followed (possibly after some time) by b.

Given a dependency between two activities, we present a couple of rules which, combined, indicate if there is handover of role between the two activities. Specifically, the combination of rules indicates a measure of the expectation of handover between roles.

14.2.1 Rule for Strong No Handover

The first rule is used to identify the absence of handover of role. In this case, given the multiset $\mathscr{U}_{a \to b}$ for a dependency between two activities $a \to b$, the idea is to check if there are couples $(u, v) \in \mathscr{U}_{a \to b}$ such that $u = v$. If this is the case, it means that there is an originator performing both a and b. As stated previously, we assume that one person hardly holds more than one role; thereby there is no handover of role between subsequent activities performed by the same originator.

14.2.2 Rule for No Handover

The previous rule applies only on very specific situations. More generally, given a dependency $a \to b \in \mathscr{D}$, if the two sets of originators are equal, i.e. $U_a = U_b$, we assume there is no handover of role. This rule can be seen as a weaker version of the previous one: there are originators interchangeably performing a and b. On the contrary, if $\mathscr{U}_a \cap \mathscr{U}_b = \emptyset$ then, each activity has a disjoint set of originators and this is the basic assumption to have handover of role between a and b.

In typical business scenarios, however, it is very common to have border-line situations, and that is why a "boolean-valued" approach is not feasible. In the following, we propose a metric to capture the degree of handover of role between two activities.

14.2.3 Degree of No Handover of Roles

Given a process P, a dependency $a \to b \in \mathscr{D}(P)$, and the respective multisets $\mathscr{U}_{a \to b}^a$, $\mathscr{U}_{a \to b}^b$ and $\mathscr{U}_{a \to b}$, it is possible to define the degree of no handover of role w_{ab}, which captures the rules above mentioned:

$$w_{ab}(L) = \frac{|\mathscr{U}_{a \to b}^a(L) \cap \mathscr{U}_{a \to b}^b(L)| + |\sigma_=(\mathscr{U}_{a \to b}(L))|}{|\mathscr{U}_{a \to b}^a(L)| + |\mathscr{U}_{a \to b}^b(L)|}, \tag{14.1}$$

The numerator of this equation considers the intersection of the two multisets of originators (to model no handover) plus the number of originators that perform both activities a and b (to model strong no handover). These weights are divided by the sum of the sizes of the two multisets of originators.

By definition, Eq. 14.1 identifies the absence of handover of role. Specifically, it assumes values in the closed interval $[0, 1]$, where 1 indicates there is no handover of roles and 0 indicates handover. Since the ideal case (i.e., completely disjoint sets of originators for each role) is very unlikely, we propose to use a threshold τ^w on the value w_{ab}. If $w_{ab} > \tau^w$, then there is no handover of roles; otherwise the handover occurs. A partition of the activities can then be obtained by removing from the process model all the dependencies which corresponds to handovers: connected activities are in the same element of the partition (see Fig. 14.2).

Example 14.1. Given a process P, a log L, and the dependency $a \rightarrow b \in \mathscr{D}(P)$, assume that:

- $\mathscr{U}^a_{a \rightarrow b}(L) = \{u^1_1, u^1_2, u^1_3\}$,
- $\mathscr{U}^b_{a \rightarrow b}(L) = \{u^1_1, u^1_2, u^1_3\}$, and
- $\mathscr{U}_{a \rightarrow b}(L) = \{(u_1, u_1)^1, (u_2, u_2)^1, (u_3, u_3)^1\}$.

The value $w_{ab}(L) = 1$ strongly indicates there is no handover of role in this case. In fact, as the set $\mathscr{U}_{a \rightarrow b}(L)$ suggests, the same originator is observed performing both a and b several times.

Example 14.2. Let's now consider a scenario completely different from Example 14.1. Given a process P, a log L, and the dependency $a \rightarrow b \in \mathscr{D}(P)$, assume that:

- $\mathscr{U}^a_{a \rightarrow b}(L) = \{u^1_1, u^1_2, u^1_3\}$,
- $\mathscr{U}^b_{a \rightarrow b}(L) = \{u^1_4, u^1_5, u^1_6\}$, and
- $\mathscr{U}_{a \rightarrow b}(L) = \{(u_1, u_4)^1, (u_2, u_5)^1, (u_3, u_6)^1\}$.

The value $w_{ab}(L) = 0$ strongly indicates the presence of handover of role. It is evident that the two sets of originators do not share any person and, based on our assumptions, this is a symptom of handover.

Example 14.3. Consider now a third example, in the middle between Example 14.1 and Example 14.2. Given a process P, a log L, and the dependency $a \rightarrow b \in \mathscr{D}(P)$, assume that:

- $\mathscr{U}^a_{a \rightarrow b}(L) = \{u^1_1, u^1_2, u^1_3\}$,
- $\mathscr{U}^b_{a \rightarrow b}(L) = \{u^1_1, u^1_2, u^1_4\}$, and
- $\mathscr{U}_{a \rightarrow b}(L) = \{(u_1, u_1)^1, (u_2, u_4)^1, (u_3, u_2)^1\}$.

In this case, $w_{ab}(L) = 0.5$ so there is no clear handover. Looking at the originator sets, u_1 performs subsequently a and then b, in one case. Moreover, u_2 is observed performing both a and b but not on the same process instance. In this example, it turns out to be fundamental the value of the threshold τ^w, in order to decide if handover of role occurs.

14.2.4 Merging Roles

As mentioned in the introductory part, the approach presented in this context is based on two steps: the first step identifies handover of roles (through the metric w_{ab} and the threshold ρ^w), which induces a partition of activities, i.e. roles. Clearly, this way of performing the partitioning is too aggressive: if the control-flow "comes back" to roles already discovered, the handover does not entail the creation of a new role. The aim of the second step is to merge partitions that are supposed to represent the same role. Given a process P and a log L, the first step generates a partitioning $\mathbf{R}$ of the activities. In order to merge some roles, we propose a metric which returns the merging degree of two partitions. Given two roles $R_i, R_j \in \mathbf{R}$:

$$\rho_{R_i R_j}(L) = \frac{2|\mathscr{U}_{R_i}(L) \cap \mathscr{U}_{R_j}(L)|}{|\mathscr{U}_{R_i}(L)| + |\mathscr{U}_{R_j}(L)|}. \tag{14.2}$$

The basic idea of this metric is the same as presented in Eq. 14.1, i.e., to measure the amount of shared originators between the two roles. This metric produces values on the closed interval $[0, 1]$ and, if activities of the two partitions are performed by the same originators, the value of the metric is 1: therefore the two roles are supposed to be the same and merged. If the roles have no common originators, then the value of ρ is 0 and the roles are kept separated.

Due to the blurry situations that are likely in reality, a threshold τ^ρ is employed: if $\rho_{R_i R_j}(L) > \tau^\rho$ then R_i should be merged with R_j; otherwise they are considered distinct roles.

14.3 Algorithm Description

In this section, we are going to provide some algorithmic details concerning the two previously described steps. We will do this with the help of the process described in Fig. 14.1(a). Moreover, we will suggest an algorithm to generate all "plausible" partitions of activities (sets of candidate roles).

14.3.1 Step 1: Handover of Roles Identification

The first step of our approach consists in the identification of the partitions induced by every handover of role. Please note that, in our context, an handover of role may occur only when the work passes from one activity to another (i.e., dependencies between activities of the process).

To achieve our goal, given a process P, the algorithm starts by extracting all the dependencies $\mathscr{D}(P)$. After that, every dependency is weighted using Eq. 14.1 (the result is reported in Fig. 14.2(a)). At this point, we apply a threshold τ^w. Specifically, we consider a particular dependency as handover of role only if its weight is less or equal to τ^w. Every time an handover is observed, the corresponding dependency is removed from the process.

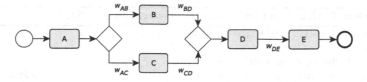

(a) Weighted dependencies.

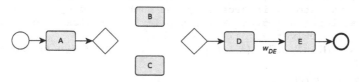

(b) Removed dependencies associated to handover of roles.

Fig. 14.2 Process model of Fig. 14.1(a) with weights associated to every dependency (*top*), and after the dependencies associated to handover of roles are removed (*bottom*). Activities are thus partitioned into the subsets $\{A\}$, $\{B\}$, $\{C\}$, $\{D, E\}$.

Let's consider again the example process of Fig. 14.1(a) and the weights of Fig. 14.2(a). Let's assume $w_{ab} \leq \tau^w$, $w_{ac} \leq \tau^w$, $w_{bd} \leq \tau^w$, $w_{cd} \leq \tau^w$ and $w_{de} > \tau^w$. Figure 14.2(b) reports the process obtained after handover of roles have been removed.

At the end of the first step, four roles have been identified: $\{A\}$, $\{B\}$, $\{C\}$, and $\{D, E\}$. These roles correspond to the activities of the connected components [41] of Fig. 14.2(b).

14.3.2 Step 2: Roles Aggregation

As previously stated, the first step of the approach identifies roles which may be too fine grained. For example, in Fig. 14.2(b) each connected component represents a role, however, as Fig. 14.1(b) shows, we actually want A in the same role of D and E, and we want B together with C. In this step, we use Eq. 14.2 to evaluate if any couple of roles may be merged.

Algorithm 4 proposes the pseudocode of the procedure used in the second phase. It requires, as input, a log L, a set of roles (i.e., a partitioning of activities) $\mathbf{R}$ and a value for the threshold τ^ρ. First of all, the algorithm finds the best pairs of roles that can be merged (line 3), i.e., pairs with maximal ρ. If the best value of ρ is above the threshold τ^ρ, it means that it is possible to merge two roles. However, several pairs may have the same maximal ρ. The criterion to select just one pair is to consider the roles that maximize the number of affected originators. If there are several pairs with identical ρ values and number of affected originators, we choose the pair that

---

Algorithm 4. Algorithm to perform roles aggregation (i.e. "Step 2")

Input: Log L; a set of roles $\mathbf{R}$; and threshold $\tau^\rho \in [0, 1]$

1 **repeat**
2     $\rho_{max} \leftarrow \max_{(R_i, R_j) \in \mathbf{R} \times \mathbf{R}} \rho_{R_i R_j}(L)$
3     $R_{\rho_{max}} \leftarrow \arg\max_{(R_i, R_j) \in \mathbf{R} \times \mathbf{R}} \rho_{R_i R_j}(L)$ `/* Maximals */`
4   **if** $\rho_{max} \geq \tau^\rho$ **then**
5   Choose $(R_i, R_j) \in R_{\rho_{max}}$ `/* Selection is performed considering`
                                     `the couple that maximizes the number of merged`
                                     `originators, if necessary the number of merged`
                                     `activities and, finally, the lexicographical order of`
                                     `role activities. */`
6        $\mathbf{R} \leftarrow (\mathbf{R} \setminus \{R_i, R_j\}) \cup \{R_i \cup R_j\}$ `/* Merge` R_i `and` R_j `*/`
7   **end**
8 **until** *no merge is performed*
9 **return R**

---

maximizes the number of merged activities. If we still have more than one pair, we just pick the first pair according to lexicographical order of contained activities (line 5). The two selected candidate roles are then merged. The same procedure is repeated until no more roles are merged (line 8), i.e., there is no pair with value of ρ above the threshold τ^ρ. Finally, the modified set of roles is returned (line 9).

14.3.3 Generation of Candidate Solutions

The approach, as presented so far, requires the configuration of two thresholds, i.e. τ^w and τ^ρ. Little variations in configuration of these parameters may lead to very different roles. To tackle this problem, we thought it might be interesting to extract all the significant partitioning and propose them to the user. Given the set of tasks A, the number of possible partitions is identified by the Bell number [18]. This quantity, given n as the size of the set, is recursively defined as:

$$B(n) = \sum_{t=0}^{n-1} \binom{n-1}{t} B(t)$$

Figure 14.3 presents the explosion of the number of possible partitioning, given the number of elements of a set.

By construction, the proposed approach requires two parameters: τ^w and τ^ρ. The values of these two thresholds are required to be in the interval $[0, 1]$; however, it can be seen that only a finite number of values produces different results (similarly to the problem tackled in Sect. 12.4).

As example, given τ^w, we can use it to remove edges from the original process. Since the number of edges of a process is finite, there is a finite number of values

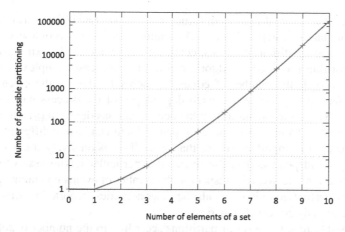

Fig. 14.3 Representation of the growth of the possible partitioning number, given the number of elements of a set.

Algorithm 5. Complete algorithm to automatically find all different partitioning of activities, given a log, and a process model.

Input: Process P; and a log L

1 $S \leftarrow \emptyset$ `/* Set of final solutions */`

2 $T^w \leftarrow \{w_{ab}(L) \mid a \rightarrow b \in \mathscr{D}(P)\}$

3 **forall the** $\tau^w \in T^w$ **do**

4 Copy the process P in P'

   `/* Step 1`   `*/`

5 **forall the** $a \rightarrow b \in \mathscr{D}(P)$ **do**

6   **if** $w_{ab}(L) \leq \tau^w$ **then**

7   Remove dependency $a \rightarrow b$ from P'

8   **end**

9 **end**

10 $R \leftarrow$ set of activities in connected components of P'

11 $T^\rho \leftarrow \{\rho_{R_i R_j}(L) \mid R_i, R_j \in R\}$

12 **forall the** $\tau^\rho \in T^\rho$ **do**

    `/* Step 2`   `*/`

13   $R_{final} \leftarrow$ **Roles Merger** (L, R, τ^ρ) `/* See Algorithm 4 */`

14   $S \leftarrow S \cup \{R_{final}\}$ `/* Consider the new solution */`

15 **end**

16 **end**

17 **return** S

---

of τ^w that splits activities of the process. The same observation can be applied to enumerate the possible values of τ^ρ.

The algorithm described in Algorithm 5 proposes an approach which automatically extracts all the significant configurations of τ^w and τ^ρ and returns such set of solutions. Specifically, line 2 collects all the significant values of τ^w. All these values are used to remove the handover of roles (line 5–9). In line 11, given the partitioning just obtained, the set of all significant values for τ^ρ is generated. These are considered for the computation of step 2 (line 13). The returned result consists of a set containing all the significant partitions (with respect to the log L) that can be extracted.

The algorithm proposed in Algorithm 5 has a worst-case complexity which is $O(n^3)$, where n is the number of edges (i.e. dependencies) of the given process model. In fact, it is possible that each dependency of the process has a different weight w_{ab}. The same situation may happen when considering ρ_{AB}: it is possible to have n clusters from step 1, and each pair of them can have a different value of ρ_{AB}. However, it is important to note that, typically, n is relatively small and, more importantly, is independent from the given log. In particular, it is necessary to analyze the log (linear complexity, with respect to the number of events it contains), but this operation is performed only once: all the other activities (reported in Algorithm 5) can use the already collected statistics.

It is possible to sort the set of partitions according to the number of roles. This ordered set is then proposed to the final user. In this way, the user will be able to explore all the significant alternate definitions of roles.

14.3.4 Partition Evaluation

A possible way to evaluate the discovered partitions is to use the concept of entropy [139]. In this context, we propose our own measure. Specifically, given $\mathbf{R}$ as the current partition, i.e., set of roles (each role is a set of activities), U as the set of originators, and L as a log, we define an entropy measure of the partition as:

$$H(\mathbf{R}, L) = \sum_{u \in U} \sum_{R \in \mathbf{R}} -\frac{|L|_R^u}{|L|^u} \log_2 \left(\frac{|L|_R^u}{|L|^u} \right). \qquad (14.3)$$

Let us recall that $|L|_R^u$ is defined as the number of times that activities belonging to the role R, and performed by user u, are observed in L; and that $|L|^u$ is defined as the number of activities executed by originator u in the log L. This measure is zero if each originator is involved in one and only one role. Otherwise, the measure increases with the degree of mixture of contribution of originators to multiple roles.

14.4 Experiments

The approach just presented has been evaluated against a couple of artificial dataset. In our datasets, we already have the target partitioning (i.e. the expected roles) and, given a log, our goal is to discover those roles. To evaluate our results, we will compare the target roles with the extracted ones and we will use a measure inspired by purity [103]. Let us recall that A represents the set of activities (or tasks) of the process and that a role is a set of activities. $|R|$ is the number of activities contained in R. Given the target partition (i.e. a set of roles) $\mathbf{R}_t$ and the discovered one $\mathbf{R}_d$, our degree of similarity is defined as:

$$similarity = \frac{1}{|\mathbf{R}_d|} \sum_{R_d \in \mathbf{R}_d} \max_{R_c \in \mathbf{R}_c} \frac{2|R_d \cap R_c|}{|R_d| + |R_c|}.$$

The idea behind this formulation is that, if the partitioning discovered is equal to the target, the similarity value is 1, otherwise it decreases.

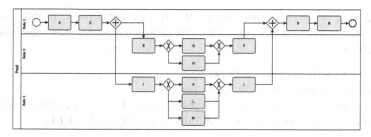

(a) Model 1.

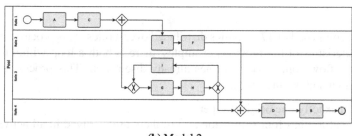

(b) Model 2.

Fig. 14.4 Process models generated for the creation of the artificial dataset.

Four artificial processes have been created (see Chap. 16). These processes, two of them shown in Fig. 14.4, have been simulated 1000 times.

Model 1

Model 1 (Fig. 14.4(a)) contains 13 activities divided over 3 roles. A peculiarity of this process is that the workflow starts with activities belonging to "Role 1" and finishes with other activities belonging to the same "Role 1". This processes have been simulated to generate five different logs:

1. one with exactly one originator per role;
2. another with exactly two originators per role;

3. the third log is similar to the second but is also includes a "jolly": an originator performing all activities;
4. the fourth log contains three originators; all of them are involved on all activities, however, each role has a "leader". Given a role, an activity is executed by its leader with probability 0.5, otherwise all other originators are equally likely;
5. the last log has 6 originators performing all the activities with a leader for each role (with the same probabilities of the previous case).

Model 2

Model 2 (Fig. 14.4(b)) is composed by 9 activities and 4 roles. In this case, the process also has a loop of activities within "Role 3". This process has been simulated to generate 3 logs:

1. one with exactly one originator per role;
2. another with exactly two originators per role;
3. the last one with 8 originators, all of them involved in all the activities, with one "leader" per role (with same probabilities of last logs of Model 1).

Model 3

Model 3 is composed of 17 activities distributed over 4 roles. This process combines two characteristics of the previous examples: there is both a loop within the same role and the flow comes back to roles already discovered. This process has been simulated to generate three logs:

1. one with exactly one originator per role;
2. the second log has four originators, all of them are involved in all activities but each role has one leader (same probabilities of previous cases);
3. the last log is characterized by 8 originators.

Model 4

Model 4 is composed of 21 activities distributed over 4 roles. In this last case, the flow starts and finishes with activities belonging to the same roles (so there is a loop). Moreover, this loop is located between activities of the two "externals" roles, so the entire process (and therefore the roles) can be observed several times on the same trace. This process has been simulated to generate three logs:

1. one with exactly one originator per role;
2. the second logs has four originators, plus one jolly, involved in all activities;
3. the last log is characterized by 8 originators.

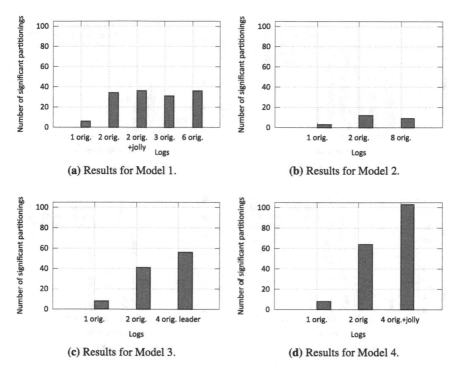

Fig. 14.5 These charts report the results, for the four models, in terms of number of significant different partitions discovered.

Results

The first results are presented in Fig. 14.5. Specifically, for each log, the number of different partitions is reported. Please note that this number is always relatively small. The worst case is observed on the fourth model, on the log with a jolly. This is what we actually expect: we have a very low number of originators (just 4) and one jolly involved indiscriminately in all activities. Moreover, the structure of the process allows having the same role appearing several times into the same trace.

Figure 14.6 proposes, for the four models, the distribution of the partitions according to the corresponding similarity measure (with respect to target roles). Concerning the logs of Model 1, all the partitions present very high similarity values, most of them concentrated on the interval $[1, 0.5]$. In the case of Model 2, most of partitions lay on the interval $[1, 0.7]$. The last two models have a modestly wider distribution of values; however, it is very important to note that in all cases the system extracts the target set of roles (i.e. there is always a partition with similarity 1).

The last result is presented in Table 14.1, whose purpose is to evaluate the entropy measure. Specifically, for each log, we ranked all partitions according to the corresponding entropy measure. After that, we verified the position of the target

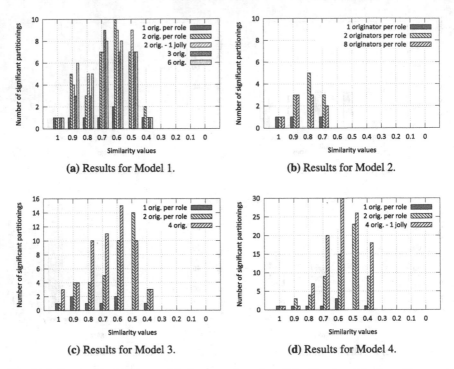

(a) Results for Model 1.

(b) Results for Model 2.

(c) Results for Model 3.

(d) Results for Model 4.

Fig. 14.6 Results, for the four models, in terms of number of significant partitioning with respect to the purity value, reported in bin of width 0.1.

partition. Results are reported in Table 14.1 and, as you can see, whenever there is no "confusion" (i.e. one originator is involved in exactly one role), the entropy measure suggests the desired partition (i.e. the target partition is in first place). Instead, when the same originator performs several roles, the confusion increases and it is harder, for our entropy measure, to correctly identify the target partition (i.e. the target partition is not in first place).

14.5 Other Approaches Dealing with Organizational Perspective

The *organizational perspective* of process mining aims at the discovery of relations among activity originators. Typically, those activities involve several approaches, such as classification of users in roles and social network analysis.

In [143], Song and van der Aalst present an exhaustive characterization of organizational mining approaches. In particular, three different types of approaches are presented: (*a*) organizational model mining; (*b*) social network analysis; and (*c*) information flows between organizational entities.

	Logs	Rank of target partition
Table 14.1 This table reports, for each log, the rank of the target partition. Ranking is based on the entropy value.	Model 1 1 originator per role	1
	2 originators per role	1
	2 originators per role – 1 jolly	4
	3 originators with leader	4
	6 originators with leader	12
	Model 2 1 originator per role	1
	2 originators per role	1
	8 originators with leader	5
	Model 3 1 originator per role	1
	2 originators per role	1
	4 originators with leader	19
	Model 4 1 originator per role	1
	2 originators per role	1
	4 originators per role – 1 jolly	30

Organizational model mining consists in grouping users with similar characteristics. This grouping operation can rely on the similarity of activities performed (task based) or on working together on the same process instance (case based).

The basic idea of social network analysis [157] is to discover how the work is handled between different originators. Several metrics are employed to point out different perspectives of the social network. Examples of such metrics are the *handover of work* (when the work is started by a user and completed by another one), *subcontracting* (when activities are performed by user a, then user b and then a again) and *working together* (users involved in the same case).

The information collected in social networks can be aggregated in order to produce *organizational entities* (such as roles or organizational unit). These entities are useful to provide insights at a higher abstraction level. Organizational entities are constructed considering a metric, and the deriving social network and aggregating nodes. These new connections can be weighted according to the weights of the originating network.

However, none of the above-mentioned works specifically addresses the problem of discovering roles in business processes.

14.6 Summary

This chapter considered the problem of extending a business process model with information about roles. Specifically, we showed that the discovery of a partitioning of activities. To achieve our goal, we took into account originators and activities they perform. Measures of handover of roles are defined and employed. Finally, we proposed an approach to automatically extract only the significant partitionings. These set of possible roles can be ranked according to an entropy measure, so that analyst may explore only first results.

With respect to the problems mentioned in Sect. 1.2, this chapter deals with problem **P-02**: exploiting as much available information as possible, *during* the actual mining phase (see Sect. 8.3). Moreover, we solved the problem considering also **P-03**: no user interaction is required, and therefore the approach is suitable also for non-expert users. Another approach that exploits additional information was described in Chap. 11.

Chapter 15
Results Interpretation and Evaluation

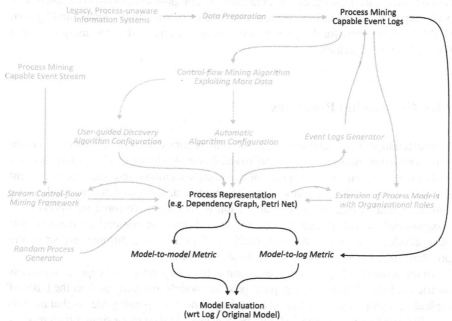

Process mining algorithms, designed for real world data, typically cope with noisy or incomplete logs via techniques that force the analyst to set the value of several parameters. Because of that, many process models corresponding to different parameters settings can be generated, and the analyst gets very easily lost in such a variety of process models. In order to have really effective algorithms, it is of paramount importance to give to the analyst the possibility to easily interpret the output of the mining.

Chapter 12 proposes a technique to automatically discretize the space of the values of the parameters, and a technique for selecting one model among all the ones that can be mined. However, presenting just a single output model could not be enough informative for the analyst (problems with the *interpretation* of the results, as mentioned

© Springer International Publishing Switzerland 2015

A. Burattin: *Process Mining Techniques in Business Environments*, LNBIP 207,
DOI 10.1007/978-3-319-17482-2_15

in Sect. 8.4); so, to solve the problem we need to find a way of presenting only a small set with the most meaningful results, so that the analyst can either point out the one that better fits the actual business context, or extract general knowledge about the business process from a set of relevant extracted models. In Sect. 1.2, this problem is reported as **P-04**.

In order to pursue this objective, it is necessary to make possible the comparison between process models, so to avoid that the analyst has to deal with too similar processes. We propose a model-to-model metric that allows the comparison between business processes, removing some of the problems which afflict other metrics already proposed in the literature. The proposed metric, in particular, transforms a given model into two sets of relations among process activities. The comparison of two models is then performed on the generated sets.

On the second part of this chapter we will propose a model-to-log metric, useful for conformance checking. In particular, we will compare a declarative process model with respect to an event log. We are also able to provide both "local" and "global healthiness" measure for the given process, which can be used by the analyst as input for further investigations.

15.1   Comparing Processes

The selection of those perspectives that should be considered relevant for comparison between two business processes in not trivial. For example, we can have two processes with the same structure (in terms of connections among activities) but different activity names. In this case, it is easy, for a human analyst, to detect the underlying similarity, while a machine will hardly be able to capture this feature unless previously programmed to do that. For this reason, several different comparison metrics have been developed in the recent past, each one focusing on a different aspect of the problem and related to a specific similarity measure.

In the context of business process mining, the first works that proposed a process metric are [46, 154]. In those papers, process models are compared on the basis of typical behaviors (expressed as an event log). The underpinning idea is that models that differ on infrequent traces should be considered much more similar than models that differ on very frequent traces. Of course, this requires the necessity of a reference execution log. In [51], the authors address the problem of detection of synonyms and homonyms that can occur when two business processes are compared. Specifically, a syntactic similarity is computed by comparing the number of characters of the activities names; linguistic similarity depends on a dictionary of terms, and structural similarity is based on the hierarchical structure of an ontology. These three similarities are combined in a weighted average. The work by Bae et al. [7] explicitly refers to process mining as one of its purposes. The authors propose to represent a process via its corresponding dependency graph, which in turn is converted into its incidence matrix. The distance between two processes is then computed as the trace of $(N_1 - N_2) \times (N_1 - N_2)^T$, where N_1 and N_2 are the process incidence matrices.

Authors of [174] present an approach for the comparison of models on the basis of their "causal footprints". A causal footprint can be seen as a collection of the essential behavioral constraints that a process model imposes. The similarity between processes is computed on the basis of their corresponding causal footprints, using the cosine similarity. Moreover, in order to avoid synonyms, a semantic similarity among function names is computed. The idea behind [48] is slightly different from the above mentioned works as long as it tries to point out the differences between two processes, so that a process analyst can understand them. Actually, this work is based on [174]. The proposed technique exploits the notion of complete trace equivalence in order to determine differences. The work by Wang et al. [181] considers only Petri nets. The basic idea is that the complete firing sequence of a Petri net might not be finite, so it is not possible to compare Petri nets in these terms. That's why the Petri net is converted into the corresponding coverability tree (guaranteed to be finite) and the comparison is performed on the principal transition sequences, created from the corresponding coverability trees. The paper [186] describes a process in terms of its "Transition Adjacency Relations" (TAR). The set of TARs describing a process is the set of pairs of activities that occur one directly after the other. The TAR set of a process is always finite, so the similarity measure is computed between the TAR sets of the two processes. The similarity measure is defined as the ratio between the cardinality of the intersection of the TARs and the cardinality of the union of them. A recent work [182] proposes to measure the consistency between business processes representing them as "behavioral profiles", which are defined as the set of strict order, exclusiveness and interleaving relations. The approach for the generation of these sets is based on Petri nets (their firing sequences) and the consistency of two processes is calculated as the amount of shared holding relations, according to a correspondence relations, that maps transition of one process into transitions of the other. Reference [93] describes a metric which takes into account five simple similarity measures, based on behavioral profiles, as the previous case. These measures are then compared using Jaccard coefficient.

15.1.1 Problem Statement and the General Approach

The first step of our approach [5] consists in converting a process model into another formalism where we can easily define a similarity measure. We think that the idea of [186], presented before, can be refined to better fit the case of business processes. In that work, a process is represented by a set of TARs. Specifically, given a Petri net P, and its set of transitions T, a TAR $\langle a, b \rangle$ (where $a, b \in T$) exists if and only if there is a trace $\sigma = t_1 t_2 t_3 \ldots t_n$ generated by P and $\exists i \in \{1, 2, \ldots, n-1\}$ such that $t_i = a$ and $t_{i+1} = b$. For example, if we consider the two processes of Fig. 15.1, they have the same TAR sets: all the possible traces generated by them always start with the transition named "A" and end with "D". In the middle, the process on the left hand side has two AND branches with the transitions "B" and "C" (so the TAR set must take into account all the possible combinations of their executions); the right hand

Fig. 15.1 Two processes
described as Petri nets that
generate the same TAR sets.
According to the work
described in [186], their
similarity would be 1, so
they would be considered
essentially as the same
process.

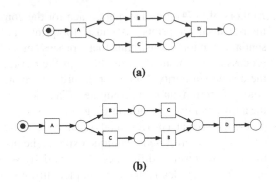

(a)

(b)

side process has two XOR branches, and they describe all the possible combinations
of the activities. Because of this peculiarity, the pairs of adjacent transitions that both
process models can generate are the same, so their similarity measure is 1 (i.e. they
describe the same process).

The main problem with this metric is that, even if from a "trace equivalence" point
of view, the two processes in Fig. 15.1 are the same (considering the two TAR sets),
from a more practical (i.e. business processes) point of view they are not: e.g., the
second process contains repeated activities and, more importantly, if activities "*B*"
and "*C*" last for a certain time period (i.e. they are not instantaneous), then it is not
the same to observe them in parallel or in (all the possible) sequences. Moreover,
there are many processes that will generate the same set of traces and a metric for
the comparison of processes should consider them as different.

Similarly to the above-mentioned works we also propose, at first, to convert a
process model from a representation hard to work with (such as Petri net or a Heuristics Net), into another one easier to handle; then, the real comparison will be performed on these new representations. However, in our case the model is transformed
into two sets of relations instead of one. In this way, the comparison is performed by
combining the results obtained by the comparison of the two sets individually.

15.1.2 Process Representation

As previously mentioned, we first have to convert the process into two sets: one set of
relations between activities that *must* occur, and another set of relations that *cannot*
occur. For example, consider the process in Fig. 15.2(a), where a representation of
the process as a Petri net is given. That is a simple process that contains a parallel
split in the middle. In Fig. 15.2(b), the same process is given but it is represented as
a dependency graph.

In order to better understand the representation of business processes we are
introducing, it is necessary to give the definition of *workflow trace*, i.e., the sequence
of activities that are executed when a business process is followed. For example,
considering again the process in Fig. 15.2, the set of all the possible traces that can
be observed is:

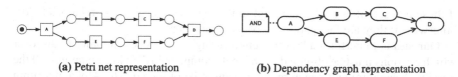

(a) Petri net representation   (b) Dependency graph representation

Fig. 15.2 An example of business process presented as a Petri net and as a dependency graph.

$$\{ABCEFD, ABECFD, ABEFCD, AEBCFD, AEBFCD, AEFBCD\}.$$

We propose to represent these processes using two types of relations: a first set containing those relations that must hold, the second set containing those relations that cannot hold. Specifically, we consider relations ($A > B$ and $A \not> B$) which have been already used by the Alpha algorithm (Sect. 5.1).

More formally, if a relation $A > B$ holds, it means that, in some workflow traces that the model can generate, activities A and B are adjacent: let W be the set of all the possible traces of a model, then there exists at least one trace $\sigma = t_1 \ldots t_n \in W$, where $t_i = A$ and $t_{i+1} = B$ for some $i \in \{1, \ldots, n-1\}$.

The other relation, $A \not> B$, is the negation of the previous one: if it holds, then, for any $\sigma = t_1 \ldots t_n \in W$, there is no i such that $t_i = A$ and $t_{i+1} = B$. It is important to note that the relations above-shown describe only local behaviors (i.e., they do not consider activities that occur far apart). Moreover, it must be noticed that our definition of $>$ is the same as the one used in [186].

These relations have been presented in [105, 162, 170] and are used by the Alpha algorithm for calculating the possible causal dependency between two activities. However, in the case considered by the cited papers, the idea is different: given a workflow log W, the Alpha algorithm finds all the $>$ relations and then, according to some predefined rules, these relations are combined to get more useful derived relations. The specific rules, mined starting from $>$, are:

1. $A \to B$, iif $A > B$ and $B \not> A$;
2. $A\#B$, iif $A \not> B$ and $B \not> A$;
3. $A\|B$, iif $A > B$ and $B > A$.

In this case, the relations $>$ and $\not>$ will be called *primitive relations*, while $\to$, # and $\|$ will be called *derived relations*. The basic ideas underpinning these three rules are:

1. if two activities are observed always adjacent and in the same order, then there should be causal dependency between them ($\to$);
2. if two activities are never seen as adjacent activities, it is possible that they are not in any causal dependency (#);
3. if two activities are observed in no specific order, it is possible that they are in parallel branches ($\|$).

Starting from these definitions, it is clear that, given two activities contained in a log, at most one derived relation ($\to$, # and $\|$) can hold between them. In particular, if these two activities appear adjacent in the log, then one of these relations holds; otherwise, if they are far apart, none of the relations hold.

Our idea is to perform a "reverse engineering" of a process in order to discover which relations must be observed in an ideal "complete log" (a log containing all the possible behaviors) and which relations cannot be observed. The Alpha algorithm describes how to mine a workflow log to extract sets of holding relations, which will be then combined and converted into a Petri net. The reverse approach can be applied too, although it is less intuitive. So, our idea is to convert a Petri net into two sets: one with $>$ and the other with $\not>$ relations.

To further understand our approach, it is useful to point out the main differences with respect to the Alpha algorithm. Considering Fig. 15.3, filled lines represent what the Alpha algorithm does: starting from the log (i.e. the set of traces), it extracts the primitive relations which are then converted into derived relations and finally into a Petri net model. In our approach, that procedure is reversed and is represented with dotted lines: starting from a given model (Petri net or dependency graph, or any other process model), the derived relations are extracted and then converted into primitive ones; the comparison between business process models is actually performed at this level.

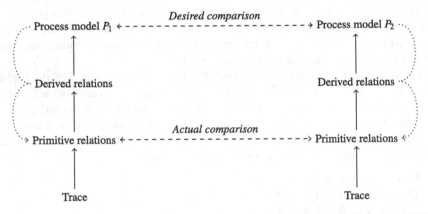

Fig. 15.3 Representation of the space where the comparison between processes is performed. The filled lines represent the steps that are performed by the Alpha algorithm. The dotted lines represent the conversion of the process into sets of primitive relations, as presented in this work.

Note that, since the naive comparison of trace equivalence is not feasible (in case of loops, the generation of the trace could never stop), we decided to analyze a model (e.g. a Petri net or a Heuristics net) and see which relations can possibly be derived. Given the set of derived relations for a model, these will be converted into two sets of positive and negative relations.

The main difference with other approaches in the literature (e.g., [182, 186]), is that our approach can be applied on every modeling language and not only Petri net or Workflow net. This is why our approach cannot rely on Petri net specific notions (such as firing sequence). We prefer to just analyze the structure of the process from a "topological" point of view. In order to face this challenge, we decided to consider a process in terms of composition of well known patterns. Right now, a small but very expressive set of "workflow patterns" [133] are taken into account. These patterns are the ones presented in Fig. 15.4.

When a model is analyzed, these derived relations are extracted:

- a sequence of two activities A and B (Fig. 15.4(a)), will generate a relation $A \rightarrow B$;
- every time an XOR split is observed (Fig. 15.4(d)) and activities A, B and C are involved, the following rules can be extracted: $A \rightarrow B$, $A \rightarrow C$ and $B\#C$; a similar approach can handle the XOR join (Fig. 15.4(e)), generating a similar set of relations: $D \rightarrow F$, $E \rightarrow F$, $D\#E$;
- every time an AND split is observed and activities A, B and C are involved (Fig. 15.4(b)) the following rules can be extracted: $A \rightarrow B$, $A \rightarrow C$ and $B\|C$; a similar approach can handle the AND join (Fig. 15.4(c)), generating a similar set of relations: $D \rightarrow F$, $E \rightarrow F$, $D\|E$.

For the case of dependency graphs, this approach is formalized in Algorithm 6: the basic idea is that, given two activities A and B, directly connected with an edge, the relation $A \rightarrow B$ must hold. If A has more than one outgoing or incoming edges $(C_1, \ldots, C_n)$ then the following relations will also hold: $C_1\rho C_2, \ldots, C_1\rho C_n, \ldots, C_{n-1}\rho C_n$ (where ρ is '#' if A is a XOR split/join, '$\|$' if A is an AND split/join).

Once the algorithm has completed the generation of the set of holding relations, this set can be split in two sets of positive and negative relations, according to the "derived relations" previously presented. Just to recap, we have $A \rightarrow B$ generates $A > B$ and $B \not> A$; $A\#B$ generates $A \not> B$ and $B \not> B$; and, finally, $A\|B$ generates $A > B$ and $B > A$.

Let's consider again the process P of Fig. 15.2. After the execution of the three "**foreaches**", in Algorithm 6 (so before the **return** of the last line), R will contain all the derived relations that, in the considered example, are:

$$A \rightarrow B \quad A \rightarrow E \quad B \rightarrow C \quad E \rightarrow F \quad C \rightarrow D \quad F \rightarrow D \quad B\|E \quad C\|F$$

These will be converted during the **return** operation of the algorithm into these two sets:

$$R^+(P) = \{A > B, \ A > E, \ B > C, \ E > F, \ C > D, \ F > D, \ B > E,$$
$$E > B, \ C > F, \ F > C\}$$

$$R^-(P) = \{B \not> A, \ E \not> A, \ C \not> B, \ F \not> E, \ D \not> C, \ D \not> F\}$$

It is important to maintain these two sets separated because of the metric we are going to introduce on the following section.

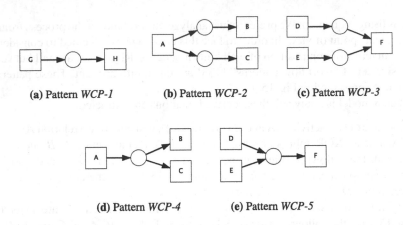

(a) Pattern *WCP-1*   **(b)** Pattern *WCP-2*   **(c)** Pattern *WCP-3*

(d) Pattern *WCP-4*   **(e)** Pattern *WCP-5*

Fig. 15.4 The basic workflow patterns that are managed by the algorithm for the conversion of a process model into set of relations. The patterns are named with the same codes of [133]. It is important to note that in *WCP-2,3,4,5* any number of branches is possible, although this picture presents only the particular case of 2 branches. Moreover, the loop is not reported here because it can be expressed in terms of XOR-split/join (*WCP-4,5*).

15.1.3 A Metric for Processes Comparison

Converting a process model into another representation is useful to compare two processes in a more easy and effective way. Here we propose a procedure to use the previously defined representations to obtain a principled metric. Specifically, given two processes P_1 and P_2, expressed in terms of positive and negative constraints: $P_1 = (R^+, R^-)$ and $P_2 = (R^+, R^-)$ they are compared according to the amount of shared "required" and "prohibited" behaviors. A possible way to compare these values is the Jaccard similarity J and the corresponding distance J_δ, that is defined in [126], between two sets, as:

$$ J(A, B) = \frac{|A \cap B|}{|A \cup B|} \qquad J_\delta(A, B) = 1 - J(A, B) = \frac{|A \cup B| - |A \cap B|}{|A \cup B|} $$

For example, it is proven that Jaccard is actually a distance measure over sets (so it is not-negative, symmetric and satisfies the identity of indiscernibles and the triangle inequality).

Our new metric is built considering the convex combination of the Jaccard distance for the set of positive and negative relations of two processes:

$$ d(P_1, P_2) = \alpha J_\delta \left(R^+(P_1), R^+(P_2) \right) + (1 - \alpha) J_\delta \left(R^-(P_1), R^-(P_2) \right) $$

where $0 \le \alpha \le 1$ is a weighting factor that allows the user to calibrate the importance of the positive and negative relations. Since this metric is defined as a linear combination of distances (J_δ), it is a distance itself. It is important that the given measure is

---

Algorithm 6. Conversion of a dependency graph into sets of relations.

Input: $G = (V, E)$: process as a dependency graph
$\mathcal{T} : V \rightarrow \{XOR\ split, XOR\ join, AND\ split, AND\ join\}$
1 R: set of holding relations
2 **foreach** $(v_1, v_2) \in E$ **do**
3 $\quad |\quad R = R \cup \{v_1 \rightarrow v_2\}$
4 **end**

5 **foreach** $v \in V, \ X = \{u \in V \mid (v, u) \in E\}$ **do**
6 $\quad |\quad$ **foreach** $(u_1, u_2) \in X \times X$ such that $u_1 \neq u_2$ **do**
7 $\quad |\quad |\quad$ **if** $\mathcal{T}(v)$ is *XOR split* **then**
8 $\quad |\quad |\quad |\quad R = R \cup \{u_1 \# u_2\}$
9 $\quad |\quad |\quad$ **else if** $\mathcal{T}(v)$ is *AND split* **then**
10 $\quad |\quad |\quad |\quad R = R \cup \{u_1 \| u_2\}$
11 $\quad |\quad |\quad$ **end**
12 $\quad |\quad$ **end**
13 **end**
14 **foreach** $v \in V, \ X = \{u \in V \mid (u, v) \in E\}$ **do**
15 $\quad |\quad$ **foreach** $(u_1, u_2) \in X \times X$ such that $u_1 \neq u_2$ **do**
16 $\quad |\quad |\quad$ **if** $\mathcal{T}(v)$ is *XOR join* **then**
17 $\quad |\quad |\quad |\quad R = R \cup \{u_1 \# u_2\}$
18 $\quad |\quad |\quad$ **else if** $\mathcal{T}(v)$ is *AND join* **then**
19 $\quad |\quad |\quad |\quad R = R \cup \{u_1 \| u_2\}$
20 $\quad |\quad |\quad$ **end**
21 $\quad |\quad$ **end**
22 **end**
23 **return** *convertRelations(R)*

---

actually a metric, because the final aim of this approach is doing clustering on those business processes.

It is important to note that there are couples of relations that are not "allowed" at the same time, otherwise the process is ill-defined and shows problematic behaviors, e.g. deadlocks[1]. Incompatible couples are defined as follows:

- if $A \rightarrow B$ holds then $A \| B$, $B \| A$, $A \# B$, $B \# A$, $B \rightarrow A$ are not allowed;
- if $A \| B$ holds then $A \# B$, $B \# A$, $A \rightarrow B$, $B \rightarrow A$, $B \| A$ are not allowed;
- if $A \# B$ holds then $A \| B$, $B \| A$, $A \rightarrow B$, $B \rightarrow A$, $B \# A$ are not allowed.

Similarly, considering primitive relations, if $A > B$ holds then $A \not> B$ represents an inconsistency, so that this behavior should not be allowed.

Theorem 15.1. *Two processes, composed of different patterns, that do not contain duplicated activities and that do not have contradictions into their set of relations (either derived or primitive), have distance measure greater than 0.*

---

[1] It must be stressed that a process may be ill-defined even if no such couples of relations are present at the same time.

Proof. Since the distance measure is calculated on the basis of the two sets of primitive relations, two processes $P_1 = (R_{P_1}^+, R_{P_1}^-)$ and $P_2 = (R_{P_2}^+, R_{P_2}^-)$ have a distance measure $d(P_1, P_2) > 0$ iff the sets $R_{P_1}^+$, $R_{P_2}^+$ and $R_{P_1}^-$, $R_{P_2}^-$ are not pairwise equal. The two sets R^+ and R^- are generated starting from the derived relations, and these are created starting from the patterns observed. If we assume that two processes are made of different patterns, they will generate different sets of derived relations and thus different sets of primitive relations. This is going to generate a distance measure, for the two processes that is greater than 0.

Since the sets of relations are generated without looking at the set of traces, but just starting from the local structure of the process model, if it is not sound (considering the Petri net notion of soundness [145]) it is possible to observe "contradictions".

There is another important aspect that needs to be pointed out: in the case of contradictions, there may be an unexpected behavior of the proposed metric. For example, in Fig. 15.5, the two processes are "structurally different", but have distance measure equals to 0. This is due to the contradictions contained in the set of primitive relations that are generated, because of the contradictions on the derived relations (in both processes $B\|C$ and $B\#C$ hold at the same time). More generally, we have that two different processes have distance measure equals to 0 when their differences results in contradictions.

Fig. 15.5 Two processes that are different and contain contradictions in their corresponding set of relations: they have distance measure equals to 0.

Consider the three processes of Fig. 15.1(a), (b) and 15.2. Table 15.1 shows the values of the TAR metric [186], compared with the ones of the metric proposed in this work, with different values of its parameter α. Note that, when $\alpha = 1$ then only the positive relations are considered; when $\alpha = 0$, only negative relations are taken into account; and, when $\alpha = 0.5$, the two cases are equally balanced. Moreover, in the situation presented here, the TAR metric and the metric of this work (with $\alpha = 1$) are equal but, generally, they are not (when there is some concurrent behavior, TAR metric adds relations with all the other activities in the other branches, whereas our metric adds only local relations with the first activities of the branches).

This procedure has been implemented and tested as discussed in Chap. 13.

Table 15.1 Values of the metrics comparing three process models presented in this work. The metric proposed here is presented with 3 values of its α parameter.

	Figure 15.1(a)–(b)	Figures 15.1(a)–15.2	Figures 15.1(b)–15.2
TAR set [186]	0	0.82	0.82
Our metric, $\alpha = 1$	0	0.77	0.77
Our metric, $\alpha = 0.5$	0.165	0.76	0.71
Our metric, $\alpha = 0$	0.33	0.75	0.66

15.2 A-Posteriori Analysis of Declarative Processes

The metric proposed in the previous section is a model-to-model metric: it aims at comparing two process models. However, for conformance checking and evaluation, we may need to analyze whether the observed behavior matches the modeled behavior. In such settings, it is often desirable to specify the expected behavior in terms of a *declarative* process model rather than of a detailed procedural one. Unfortunately, declarative models do not have an explicit notion of *state*, thus making it more difficult to pinpoint deviations and to explain and quantify discrepancies.

This section focuses on providing high-quality and understandable diagnostics measures [22]. The notion of *activation* plays a key role in determining the effect of individual events on a given constraint. Using this notion, we are able to show cause-and-effect relations and to measure the healthiness of a process.

15.2.1 Declare

Declarative languages can be fruitfully applied in the context of process discovery [34, 98, 101] and compliance checking [9, 47, 90, 100]. In [156], the authors introduce an LTL-based declarative process modeling language called *Declare*. Declare is characterized by a user-friendly graphical representation with formal semantics grounded in *Linear Temporal Logic* (LTL). A Declare model is a set of Declare constraints, which are defined as instantiations of Declare templates. Templates are abstract entities that define parameterized classes of properties.

Declare is grounded in LTL [124] with a finite-trace semantics. For instance, a constraint like the *response* constraint in Fig. 2.7 can be formally represented using LTL and, in particular, it can be written as $\Box(C \Rightarrow \Diamond S)$, that means "whenever activity *Create Questionnaire* is executed, eventually activity *Send Questionnaire* is executed". In a formula like this, it is possible to find traditional logical operators (e.g., implication $\Rightarrow$), but also temporal operators characteristic of LTL (e.g., always $\Box$ and eventually $\Diamond$). In general, using the LTL language makes possible to express constraints relating activities (atoms) through logical operators or temporal operators.

The logical operators are: implication ($\Rightarrow$), conjunction ($\land$), disjunction ($\lor$) and negation ($\neg$). The main temporal operators are: *always* ($\Box p$, in every future state p holds), *eventually* ($\Diamond p$, in some future state p holds), *next* ($\bigcirc p$, in the next state p holds) and *until* ($p \sqcup q$, p holds until q holds).

LTL constraints are not very readable for non-experts. Therefore, Declare provides an intuitive graphical front-end for LTL formulas. The LTL back-end of Declare allows us to verify Declare constraints and Declare models, i.e., sets of Declare constraints. Table 15.2 presents some Declare relations, with the corresponding LTL constraints and the graphical representation of the Declare language.

For instance, a Declare constraint can be verified on a log by translating its LTL semantics into a finite state automaton [63] that we call *constraint automaton*.

Table 15.2 Semantics of Declare constraints, with the graphical representation.

Name	Constraint	Declare Representation
Relation Templates		
responded existence(A,B)	$\Diamond A \Rightarrow \Diamond B$	A ●———— B
co-existence(A,B)	$\Diamond A \Leftrightarrow \Diamond B$	A ●———● B
response(A,B)	$\Box(A \Rightarrow \Diamond B)$	A ●———▶ B
precedence(A,B)	$(\neg B \sqcup A) \vee \Box(\neg B)$	A ——▶● B
succession(A,B)	response$(A,B) \wedge$ precedence(A,B)	A ●—▶● B
alternate response(A,B)	$\Box(A \Rightarrow \bigcirc(\neg A \sqcup B))$	A ●———▶ B
alternate precedence(A,B)	prec.$(A,B) \wedge \Box(B \Rightarrow \bigcirc(\text{prec.}(A,B)))$	A ●———▶ B
alternate succession(A,B)	alt. response$(A,B) \wedge$ alt. precedence(A,B)	A ●———▶ B
chain response(A,B)	$\Box(A \Rightarrow \bigcirc B)$	A ●———▶ B
chain precedence(A,B)	$\Box(\bigcirc B \Rightarrow A)$	A ————▶ B
chain succession(A,B)	$\Box(A \Leftrightarrow \bigcirc B)$	A ●———▶ B
Negative Relation Templates		
not co-existence(A,B)	$\neg(\Diamond A \wedge \Diamond B)$	A ●—‖— B
not succession(A,B)	$\Box(A \Rightarrow \neg(\Diamond B))$	A ●—‖▶ B
not chain succession(A,B)	$\Box(A \Rightarrow \bigcirc(\neg B))$	A ●—▮▶ B

Figure 15.6 depicts the constraint automata for the *response* constraint, the *alternate response* constraint and the *not co-existence* constraint. In all three cases, state 0 is the initial state and accepting states are indicated using a double outline. A transition is labeled with the activity triggering it. As well as positive labels, we also have negative labels (e.g., $\neg L$ for state 0 of the *not co-existence* constraint). This indicates that we can follow the transition for any event not mentioned (e.g., we can execute event C from state 0 of the *not co-existence* automaton and remain in the same state). This allows us to use the same automaton regardless of the input language. A constraint automaton *accepts a trace* (i.e., the LTL formula holds) if and only if there exists a corresponding path that starts in the initial state and ends in an accepting state.

15.2.2 An Approach for A-Posteriori Analysis

When analyzing the conformance of a process with respect to a set of constraints, it is important to note that constraints can be vacuously satisfied. Considering again the example of Fig. 2.7, if *Create Questionnaire* never occurs, then the *response* constraint holds trivially. This is commonly referred to as *vacuous satisfaction*. In this context, we will start from the existing notion of vacuity detection [10] and

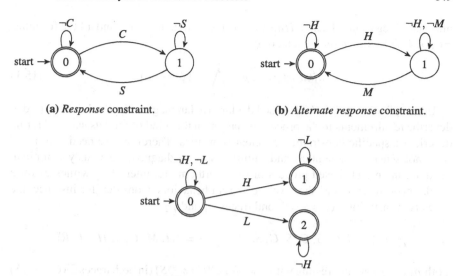

(a) *Response* constraint.   **(b)** *Alternate response* constraint.

(c) *Not co-existence* constraint.

Fig. 15.6 Automata for the *response*, *alternate response* and *not co-existence* constraints in our running example.

we will propose an approach for evaluating the "degree of adherence" of a process trace with respect to a Declare model. In particular, we will introduce the notion of *healthiness* of a trace that is, in turn, based on the concept of *activation* of a Declare constraint.

Vacuity Detection in Declare

In [101], the authors introduce for the first time the concept of vacuity detection for Declare constraints. As just stated, consider, for instance, the *response* constraint in Fig. 2.7. This constraint is satisfied when a questionnaire is created and is (eventually) sent. However, this constraint is also satisfied in cases where the questionnaire is not created at all. In this latter case, we say that the constraint is *vacuously satisfied*. Cases where a constraint is *not*-vacuously satisfied are called *interesting witnesses* for that constraint.

Authors of [94] introduce an approach for vacuity detection in temporal model checking for LTL; they provide a method for extending an LTL formula φ to a new formula $witness(\varphi)$ that, when satisfied, ensures that the original formula φ is not-vacuously satisfied. In particular, $witness(\varphi)$ is generated by considering that a path π satisfies φ not-vacuously (and then is an interesting witness for φ), if π satisfies φ and π satisfies a set of additional conditions, that guarantee that every subformula of φ does really affect the truth value of φ in π. We call these conditions *vacuity detection conditions* of φ. They correspond to the formulas $\neg\varphi[\psi \leftarrow \bot]$, where, for all the subformulas ψ of φ, $\varphi[\psi \leftarrow \bot]$ is obtained from φ by replacing ψ by false or true, depending on whether ψ is in the scope of an even or an odd number of negations. Then, $witness(\varphi)$ is the conjunction of φ and all the formulas $\neg\varphi[\psi \leftarrow \bot]$ with ψ subformula of φ:

$$witness(\varphi) = \varphi \wedge \bigwedge \neg\varphi[\psi \leftarrow \bot]. \tag{15.1}$$

In compliance models, LTL-based declarative languages, like Declare, are used to describe requirements to the process behavior. In this kind of models, each LTL rule describes a specific constraint with clear semantics. Therefore, we need a *univocal* (i.e., not sensitive to syntax) and intuitive way to diagnose vacuously compliant behavior in an LTL-based process model. Furthermore, interesting witnesses for a Declare constraint could show very different behaviors. Consider, for instance, the *response* constraint $\Box(C \Rightarrow \Diamond S)$ and traces p_1 and p_2:

$$p_1 = \langle C, S, C, S, C, S, C, S, R \rangle \qquad p_2 = \langle H, M, C, S, H, M, R \rangle.$$

Both p_1 and p_2 are interesting witnesses for $\Box(C \Rightarrow \Diamond S)$ (in both traces $\Box(C \Rightarrow \Diamond S)$ is valid and the vacuity detection condition $\Diamond C$ is also valid). However, it is intuitive to understand that in p_1 this constraint is activated four times (because C occurs four times), whereas in p_2 it is activated only once. To solve these issues, we introduce the notion of *constraint activation*.

Roughly speaking, an activation for a constraint is an event that constrains in some way the behavior of other events and imposes some obligations on them. For instance, the occurrence of an event can require the occurrence of another event afterwards (e.g., in the *response* constraint) or beforehand (e.g., in the *precedence* constraint). When an activation occurs, these obligations can refer to the future, to the past or to both. Moreover, they can require or forbid the execution of other events.

Definition 15.1 (Subtrace). Let σ be a trace. A trace σ' is a *subtrace* of σ ($\sigma' \sqsubseteq \sigma$) if σ' can be obtained from σ by removing one or more events.

Definition 15.2 (Minimal Violating Trace). Let π be a Declare constraint and $\mathscr{A}_\pi$ the constraint automaton of π. A trace σ is a *minimal violating trace* for $\mathscr{A}_\pi$ if it is not accepted by $\mathscr{A}_\pi$ and if every subtrace of σ is accepted by $\mathscr{A}_\pi$.

Definition 15.3 (Constraint Activation). Let π be a Declare constraint and $\mathscr{A}_\pi$ the constraint automaton of π. Each *event* included in a minimal violating trace for $\mathscr{A}_\pi$ is an *activation* of π.

Consider, for instance, the automaton in Fig. 15.6(a). In this case, the minimal violating trace is $\langle C \rangle$. Therefore, the *response* constraint in our running example is activated by C. Moreover, for the automaton in Fig. 15.6(b), the minimal violating trace is $\langle H \rangle$ and, then, the *alternate response* constraint is activated by H. Finally, for the automaton in Fig. 15.6(c), the minimal violating sequences are $\langle L, H \rangle$ and $\langle H, L \rangle$. The *not co-existence* constraint is, therefore, activated by both H and L.

In Table 15.3, we indicate events that represent an activation for each Declare constraint. Note that events that represent an activation for a constraint are marked

Table 15.3 Activations of Declarative constraints.

Declare constraint	Activation events
Relation Templates	
responded existence(A, B)	A
co-existence(A, B)	A, B
response(A, B)	A
precedence(A, B)	B
succession(A, B)	A, B
alternate response(A, B)	A
alternate precedence(A, B)	B
alternate succession(A, B)	A, B
chain response(A, B)	A
chain precedence(A, B)	B
chain succession(A, B)	A, B
Negative Relation Templates	
not co-existence(A, B)	A, B
not succession(A, B)	A, B
not chain succession(A, B)	A, B

with a black dot in the graphical notation of Declare, e.g., both A and B are activations for the *succession* constraint (as visualized by the black dots).

15.2.3 An Algorithm to Discriminate Fulfillments from Violations

When a trace is compliant with respect to a constraint, every activation of that constraint leads to a fulfillment. For instance, recall the two traces:

$$p_1 = \langle C, S, C, S, C, S, C, S, R \rangle \quad p_2 = \langle H, M, C, S, H, M, R \rangle.$$

in p_1, the *response* constraint ($\Box(C \Rightarrow \Diamond S)$) is activated and fulfilled four times, whereas in p_2, the same constraint is activated and fulfilled once. Notice that, when a trace is not-compliant with respect to a constraint, an activation of a constraint can lead to a fulfillment but also to a violation (and at least one activation leads to a violation). Consider, again, the *response* constraint in our running example and the trace $p_3 = \langle C, S, C, R \rangle$. In this trace, the *response* constraint is violated. However, it is still possible to quantify the degree of adherence of this trace in terms of number of fulfillments and violations. Indeed, in this case, the *response* constraint is activated twice, but one activation leads to a fulfillment (eventually an event S occurs) and the other activation leads to a violation (S does not occur eventually). Therefore, we need a mechanism to point out that the first occurrence of C is a fulfillment and the second one is a violation.

Furthermore, if we consider trace $\langle H, H, M \rangle$ and the *alternate response* constraint in our running example, we observe that the two occurrences of H cannot co-exist, but it is impossible to understand (without further information from the user) which one is a violation and which one is a fulfillment. In this case, we say that we have a *conflict* between the two activations.

Algorithm 7. Procedure to build the activation tree

Input: σ: trace; π: constraint
Result: activation tree of σ with respect to π

1 Let T be a binary tree with root labeled with an empty subtrace
2 **forall the** $e \in \sigma$ *(explored in sequence)* **do**
3 | **forall the** *leaf* l *of* T **do**
4 | | **if** *the subtrace associated to* l *is not dead* **then**
5 | | | **if** e *is an activation for* π **then**
6 | | | | $l[left] =$ new node, subtrace of l
7 | | | | $l[right] =$ new node, subtrace of $l + e$
8 | | | **else**
9 | | | | subtrace of $l =$ subtrace of $l + e$
10 | | | **end**
11 | | **end**
12 | **end**
13 **end**
14 **return** T

In order to identify fulfillments, violations and conflicts for a constraint π in a trace σ, we present Algorithm 7 that is based on the construction of a so-called *activation tree* of σ with respect to π, where every node is labeled with a subtrace of σ. The algorithm starts from a root labeled with the empty subtrace. Then, σ is replayed and the tree is finally built in the following way:

- if the current event in σ is an activation of π, two children are appended to each leaf-node: a left child labeled with the subtrace of the parent node and a right child labeled with the same subtrace augmented with the current activation;
- if the current event in σ is not an activation of π, all leaf-nodes are augmented with the current event.

At each iteration, each subtrace in the leaf-nodes is executed on the constraint automaton $\mathscr{A}_\pi$. A node is called *dead* if the corresponding subtrace is not possible according to the automaton or all events have been explored and no accepting state has been reached. Dead nodes are not explored further and crossed-out in the diagrams. At the end of the algorithm, fulfillments, violations and conflicts can be identified by selecting, among the (not-dead) leaf-nodes, the *maximal fulfilling subtraces*.

Definition 15.4 (Maximal Subtrace). Given a set Σ of subtraces of a trace σ, a *maximal subtrace* of σ in Σ is an element $\sigma' \in \Sigma$ such that $\nexists \sigma'' \in \Sigma$ with $\sigma' \sqsubset \sigma''$.

Definition 15.5 (Maximal Fulfilling Subtrace). Given a trace σ and a constraint π, let $\overline{\Sigma}$ be the set of the subtraces of σ associated to the not-dead leaf-nodes of the activation tree of σ with respect to π. Let $M \subseteq \overline{\Sigma}$ be the set of the maximal subtraces of σ in $\overline{\Sigma}$. An element of M is called *maximal fulfilling subtrace* of σ with respect to π.

Let's consider an activation a of π in σ, and all its maximal fulfilling subtraces. Then a is:

- fulfillment, if it is included in all the maximal fulfilling subtraces;
- violation, if it is not included in any maximal fulfilling subtrace;
- conflict, if it is only included in some maximal fulfilling subtraces.

Fig. 15.7 Activation tree of trace $\langle C_{(1)}, S, C_{(2)}, R \rangle$ with respect to the *response* constraint in our running example: dead nodes are crossed out and nodes corresponding to maximal fulfilling subtraces are highlighted

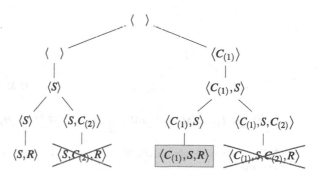

Consider, for instance, the activation tree in Fig. 15.7 of the trace:

$$\langle C_{(1)}, S, C_{(2)}, R \rangle$$

with respect to the *response* constraint in our running example (reported in Fig. 2.7). The maximal fulfilling subtrace is $\langle C_{(1)}, S, R \rangle$. We can conclude that $C_{(1)}$ is a fulfillment, whereas $C_{(2)}$ is a violation.

Figure 15.8 depicts the activation tree of trace

$$\langle H_{(1)}, M, H_{(2)}, H_{(3)}, M \rangle$$

with respect to the *alternate response* constraint in our running example. The maximal fulfilling subtraces are, in this case, $\langle H_{(1)}, M, H_{(2)}, M \rangle$ and $\langle H_{(1)}, M, H_{(3)}, M \rangle$. We can conclude that $H_{(1)}$ is a fulfillment, whereas $H_{(2)}$ and $H_{(3)}$ are conflicts.

Finally, Fig. 15.9 depicts the activation tree of trace

$$\langle H, M, L_{(1)}, L_{(2)} \rangle$$

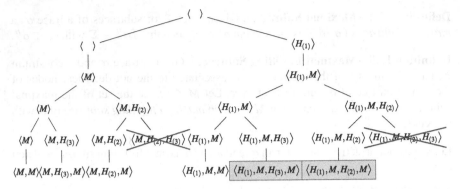

Fig. 15.8 Activation tree of trace $\langle H_{(1)}, M, H_{(2)}, H_{(3)}, M \rangle$ with respect to the *alternate response* constraint in our running example

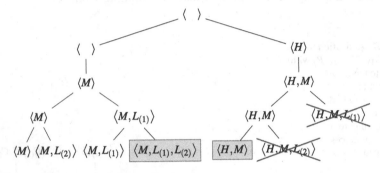

Fig. 15.9 Activation tree of trace $\langle H, M, L_{(1)}, L_{(2)} \rangle$ with respect to the *not co-existence* constraint in our example.

with respect to the *not co-existence* constraint in our running example. The maximal fulfilling subtraces are, in this case, $\langle H, M \rangle$ and $\langle M, L_{(1)}, L_{(2)} \rangle$. We can conclude that H, $L_{(1)}$ and $L_{(2)}$ are conflicts.

15.2.4 Healthiness Measures

In this section we will provide a definition of the healthiness of a trace with respect to a given constraint. Given a trace σ and constraint π, each event in the trace can be classified as activation or not based on Definition 15.3. $n_a(\sigma, \pi)$ is the number of activations of σ with respect to π. Each activation can be classified as a fulfillment, a violation, or a conflict based on the activation tree. $n_f(\sigma, \pi)$, $n_v(\sigma, \pi)$ and $n_c(\sigma, \pi)$ denote the numbers of fulfillments, violations and conflicts of σ with respect to π, respectively. $n(\sigma)$ is the number of events in σ.

Healthiness

The *healthiness* of a trace σ with respect to a constraint π is a quadruple $\mathcal{H}_\pi(\sigma) = (AS_\pi(\sigma), FR_\pi(\sigma), VR_\pi(\sigma), CR_\pi(\sigma))$, where:

1. $AS_\pi(\sigma) = 1 - \frac{n_a(\sigma,\pi)}{n(\sigma)}$ is the *activation sparsity* of σ with respect to π,
2. $FR_\pi(\sigma) = \frac{n_f(\sigma,\pi)}{n_a(\sigma,\pi)}$ is the *fulfillment ratio* of σ with respect to π,
3. $VR_\pi(\sigma) = \frac{n_v(\sigma,\pi)}{n_a(\sigma,\pi)}$ is the *violation ratio* of σ with respect to π and
4. $CR_\pi(\sigma) = \frac{n_c(\sigma,\pi)}{n_a(\sigma,\pi)}$ is the *conflict ratio* of σ with respect to π.

A trace σ is "healthy" with respect to a constraint π if the fulfillment ratio $FR_\pi(\sigma)$ is high, and both the violation ratio $VR_\pi(\sigma)$ and the conflict ratio $CR_\pi(\sigma)$ are low. If $FR_\pi(\sigma)$ is high, the activation sparsity $AS_\pi(\sigma)$ becomes a positive factor, otherwise it is symptom of unhealthiness.

It is possible to average the values of the healthiness over the traces in a log and over the constraints in a Declare model, thus obtaining aggregated views of the healthiness of a trace with respect to a Declare model, of a log with respect to a constraint and of a log with respect to a Declare model.

Likelihood of a Conflict Resolution

Consider trace $\langle H, M, L_{(1)}, L_{(2)} \rangle$ with respect to the *not co-existence* constraint in our running example. The maximal fulfilling subtraces are, in this case, $\langle H, M \rangle$ and $\langle M, L_{(1)}, L_{(2)} \rangle$ and $H, L_{(1)}$ and $L_{(2)}$ are conflicts. However, the maximal fulfilling subtraces also contain further information. In fact, H is included in one of the maximal fulfilling subtraces and $L_{(1)}$ and $L_{(2)}$ in the other one. This means that $L_{(1)}$ and $L_{(2)}$ can co-exist but both cannot co-exist with H. In this way, we can conclude that either H is a violation and $L_{(1)}$ and $L_{(2)}$ are fulfillments or, *vice versa*, H is a fulfillment and $L_{(1)}$ and $L_{(2)}$ are violations. We call the corresponding maximal fulfilling subtraces *conflict resolutions*.

The user can decide how to solve a conflict by selecting a conflict resolution. However, it is possible to provide the user with two health indicators, as support for this decision: the *local likelihood* of a conflict resolution and the *global likelihood* of a conflict resolution.

Definition 15.6 (Local Likelihood). Let σ' be a conflict resolution of a trace σ with respect to a constraint π. Let $n_a(\sigma', \pi)$ and $n_f(\sigma', \pi)$ be the number of activations and fulfillments of a conflict resolution σ', respectively. The *local likelihood* of σ' is defined as:

$$LL(\sigma') = \frac{n_f(\sigma', \pi)}{n_a(\sigma', \pi)}.$$

Note that local likelihood of a conflict resolution is a number in the interval $(0, 1)$. If we consider again the example described before, we have that $LL(\langle H, M \rangle) = \frac{1}{3}$

and $LL(\langle M, L_{(1)}, L_{(2)} \rangle) = \frac{2}{3}$. This means that, more likely, H is a violation and $L_{(1)}$ and $L_{(2)}$ are fulfillments.

In the following definition a Declare model is a pair $\mathcal{D} = (A, \Pi)$, where A is a set of activities and Π is a set of Declare constraints defined over activities in A.

Definition 15.7 (Global Likelihood). Let $\mathcal{D} = (A, \Pi)$ be a Declare model. Let σ' be a conflict resolution of a trace σ with respect to a constraint $\pi \in \Pi$. Let K be the set of the conflicting activations in σ. For each conflicting activation $a \in K$, let $\gamma(a)$ be the percentage of constraints in Π where a is a fulfillment, if a is resolved as a fulfillment in σ', or where a is a violation, if a is resolved as a violation in σ'. The *global likelihood* of σ' is defined as:

$$GL(\sigma') = \frac{\sum_a \gamma(a)}{|K|}.$$

The global likelihood of a conflict resolution is a number between 0 and 1. If we continue with the same example, we have that $GL(\langle H, M \rangle) = \frac{1}{6}$ and $LL(\langle M, L_{(1)}, L_{(2)} \rangle) = 0$. This means that, from the global point of view, more likely, H is a fulfillment and $L_{(1)}$ and $L_{(2)}$ are violations.

15.2.5 Experiments

For the a-posteriori analysis of a log with respect to a Declare model, we have implemented the *Declare Analyzer*, a plug-in of the process mining tool ProM. The plug-in takes as input a Declare model and a log, it provides detailed diagnostics, and quantifies the health of each trace (and of the whole log).

We evaluate the performance of our approach using both synthetic and real-life logs. Then, we validate our approach on a real case study in the context of the CoSeLoG project[2] involving 10 Dutch municipalities.

15.2.5.1 Scalability

In order to experimentally demonstrate the scalability of our approach, we have performed two experiments. Both these experiments have been performed on a standard laptop, with a dual-core processor with its power forced to 1.6 GHz. The presented results report the average value of the execution time over 5 runs.

In the first experiment, we verified the scalability of the technique when varying the log size. For this experiment, we have generated a set of synthetic logs by modeling the process described as running example in CPN Tools[3] and by simulating the model.

---

[2]Visit http://www.win.tue.nl/coselog for more information.

[3]The tool is freely available at http://www.cpntools.org.

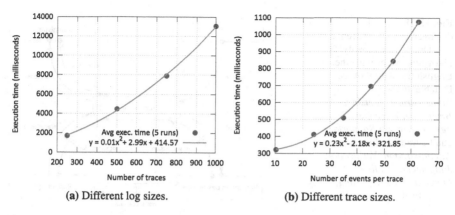

(a) Different log sizes.   **(b)** Different trace sizes.

Fig. 15.10 Execution time for varying log and trace sizes and the polynomial regression curve associated.

In particular, we used randomly generated logs including 250, 500, 750 and 1000 traces. The results are presented in Fig. 15.10(a). The plot shows that the execution time grows polynomially with the size of the logs.

In the second experiment, we evaluate the trend of the execution time with respect to the length of the traces. For this experiment, we have selected, in the CoSeLoG logs, 6 sets of traces, each composed of 10 traces of the same length. Figure 15.10(b) shows the results of this experiment. Even if the size of an activation tree is exponential in the number of activations, the execution time is polynomial in the length of the traces. Indeed, performances get worse when the number of activations is close to the number of events in a trace. However, from our experience, in practice, this case is, in general, unlikely. Specifically, the activation sparsity is in most cases high (see Table 15.4) and, therefore, the number of activations is low with respect to the number of events in a trace. This means that, from the practical point of view, the algorithm is applicable. For example, as shown in Fig. 15.10(b), processing 10 traces with 63 events requires slightly more than 1 s.

In addition, in our implementation we never needed to construct the whole activation tree of a trace. This also influences the performances of the approach. At each step of the algorithm, we keep track only of the maximal traces without building the nodes corresponding to their sub-traces. These sub-traces are identified (and evaluated) only when the original maximal trace is violated (and pruned away).

15.2.5.2 Case Study

We have validated our approach by performing various experiments using real-life event logs from the CoSeLoG project. Here, we show results for the process of handling permissions for building or renovating private houses, for which we have logs from several Dutch municipalities. For the validation reported here, we have used two logs of processes enacted by two different municipalities. We first have discovered

Fig. 15.11 Model
discovered from an event log
of a Dutch Municipality. For
clarifying, we provide the
English translation of the
Dutch activity names.
*Administratie, Toetsing,
Beslissing, Verzenden
beschikking* and *Rapportage*
can be translated with
*Administration, Verification,
Judgement, Sending
Outcomes* and *Reporting*,
respectively.

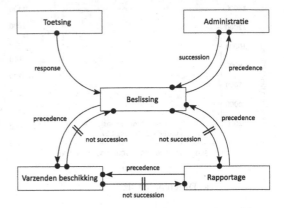

Table 15.4 Results of the analysis approach, applied to real-life event logs from the CoSeLoG
project. The table reports average activity sparsity, average violation ratio, average fulfillment ratio
and average conflicts ratio.

Constraint	Avg. act. sparsity	Avg. violat. ratio	Avg. fulfil. ratio	Avg. confli. ratio
precedence(*Rapportage,Beslissing*)	0.859	0.017	0.982	0
succession(*Administratie,Beslissing*)	0.735	0.995	0.004	0
precedence(*Rapportage,Verzenden*)	0.976	0.054	0.945	0
not succession(*Verzenden,Rapportage*)	0.731	0.000	0.999	0
response(*Toetsing,Beslissing*)	0.979	0.713	0.286	0
precedence(*Beslissing,Verzenden*)	0.976	0	0.835	0
not succession(*Verzenden,Beslissing*)	0.836	0	1	0
not succession(*Beslissing,Rapportage*)	0.614	0.000	0.995	0.004
precedence(*Beslissing,Administratie*)	0.875	0.261	0.738	0
Averages	0.842	0.245	0.754	0.000

a Declare model using an event log of one municipality using the *Declare Miner*.
This model is shown in Fig. 15.11. Then, using the *Declare Analyzer*, we have analyzed the degree of adherence of a log of the second municipality with respect to
the mined model. Analysis showed commonalities and interesting differences. From
a performance viewpoint the results were also encouraging: 481 cases with 17032
events could be replayed in 15 s.

Results are reported in Table 15.4. The fulfillment ratio is, for almost all constraints, very high and, therefore, the average fulfillment ratio over the entire Declare
model is also high (0.754). The activation sparsity of the log is, in most cases, close
to 1, indicating a low activation frequency for each constraint in the model. For the
not succession constraint between *Beslissing* and *Rapportage*, the combination of

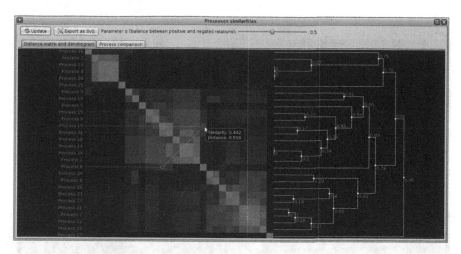

Fig. 15.12 Screenshot with the distance matrix and the dendrogram applied to some processes.

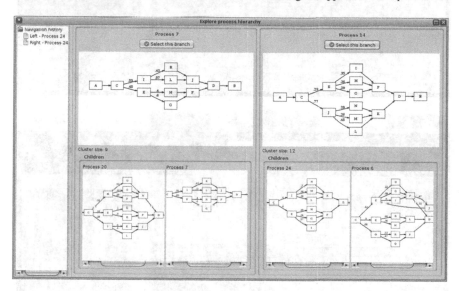

Fig. 15.13 Screenshot of the exploration procedure that allows the user to choose the most interesting process.

an under average activation sparsity with a high fulfillment ratio reveals the "good healthiness" of the log with respect to that constraint.

Nevertheless, the two municipalities execute the two processes in a slightly different manner (the average violation ratio and the conflict ratio of the log with respect to the entire Declare model are 0.245 and 0.0005 respectively). The discrepancies have mainly been detected for the *succession* constraint and for the *response* constraint in the reference model. Here, the violation ratio is high. For the *succession* constraint

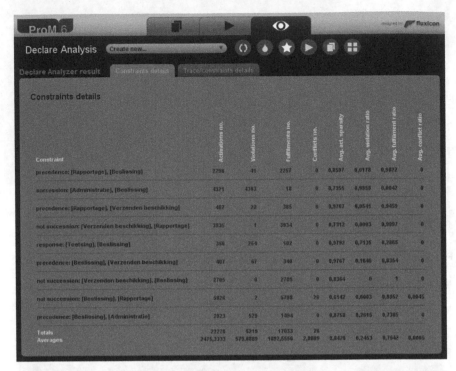

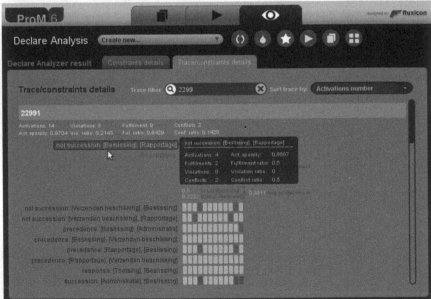

Fig. 15.14 On the left hand side there is the output of the *log view* Declare Analyzer plug-in. On the right hand side the *trace view details* is proposed.

the high violation ratio in combination with a low activation sparsity is symptom of strong unhealthiness.

15.3 Implementations

The model-to-model metric, presented in Sect. 15.1, has been implemented in the standalone application PLG (that will be presented in Sect. 16.2). In this case, as can be seen in Fig. 15.12, it is possible to create a graphical representation of the distance matrix between couples of processes. Figure 15.13 shows the procedure that allows the analyst to navigate the process clusters in order to find the most interesting process.

Concerning the a-posteriori analysis of a log with respect to a Declare model, described in Sect. 15.2, we have implemented the *Declare Analyzer* ProM plugin. This plugin takes as input a Declare model and a log and it provides detailed diagnostics, quantifying the health of each trace (and of the whole log). Figure 15.14 presets a couple of screenshots of this plugin, in particular, the log overview metrics and trace view details are proposed.

15.4 Summary

In this chapter we presented two approaches for the evaluation of process models. In particular a model-to-model and a model-to-log metrics were proposed.

The first model-to-model metric is a new approach for the comparison of business processes. This approach relies on the conversion of a process model into two sets of relations: the first contains all the local relations (only between two connected activities) that must hold; the second the relations that must not hold. These two sets are generated starting from the relations of the Alpha algorithm but, instead of starting from a log and performing abstractions to achieve some rules, the opposite way is followed: given the model, local relations (expressed in terms of behavior that is allowed in the log trace) are extracted. The proposed metric is based on the comparison of these two sets.

The second model-to-log metric is a novel approach to check the conformance of observed behavior (i.e., an event log) with respect to desired behavior modeled in a declarative manner (i.e., a Declare model). Unlike earlier approaches, thanks to this metric we are able to provide reliable diagnostics which do not depend on the underlying LTL syntax. We provided behavioral characterizations of activations, fulfillments, violations and conflicts. These can be used to provide detailed diagnostics at the event level, but can also be aggregated into health indicators such as the fulfillment ratio (fraction of activations having no problems), violation ratio and conflict ratio. Experiments showed that the approach scales well (polynomial in the size of the log and in the length of the traces). Initial experiences in a case study based on the event logs of two municipalities revealed that the diagnostics are indeed very useful and can be easily interpreted.

With respect to the problems pointed out in Sect. 1.2, this chapter deals with **P-04**: problems occurring, once the mined model is available, with the *interpretation* and the validation of the results (see Sect. 8.3).

Chapter 16
Hands-On: Obtaining Test Data

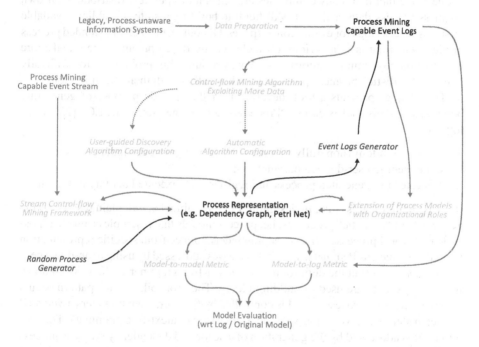

A process mining algorithm is only as good as its evaluation on actual data. Unfortunately, real life data is often not available. While one can easily synthesize data by simulating process models, event logs obtained in this way tend to show artificial patterns not found in practice, while real-life processes produce behavior not reproduced in a naïve approach.

In this chapter we will present an approach for the random generation of business processes and their execution logs. The proposed technique is based on the generation of process descriptions via a stochastic context-free grammar, whose definition is based on well-known process patterns. An algorithm for the generation of execution instances will be proposed as well.

© Springer International Publishing Switzerland 2015
A. Burattin: *Process Mining Techniques in Business Environments*, LNBIP 207,
DOI 10.1007/978-3-319-17482-2_16

16.1 A Process and Logs Generator

A critical issue concerning process mining algorithms is the evaluation of their performance, i.e., how well the reconstructed process model matches the actual process. The ideal setting to perform this evaluation entails the availability of an as-large-as-possible suite of business processes. Actually, not only business process models are required, but also one or, preferably, more logs, created according to the process they refer to. Starting from them, different process mining algorithms can be evaluated by checking the corresponding results of the mining against the original models. Figure 5.7 shows visual representation of such evaluation "cycle".

Unfortunately, it is often the case that just one (even partial) log file is available, while no clear definition of the business process that generated it is available. This is due to the fact that many companies are reluctant to publicly distribute their own business process data, and so it is difficult to build up a suite of publicly available business process logs for evaluation purposes. Of course, the lack of extended process mining benchmarks is a serious obstacle for the development of new and more effective process mining algorithms. A way around this problem is to artificially generate "realistic" business process models together with their execution logs.

This chapter presents a tool, developed for the specific purpose of generating benchmarks. This tool is called "Processes Logs Generator" (or PLG) [25]. It allows to:

1. generate random (hopefully "realistic") business process models (according to some specific user-defined parameters);
2. "execute" the generated process and register each executed activity in log files.

It is important to stress that, in designing PLG, we aimed at two main results: *(i)* the generation of "realistic" processes, i.e. process models that resemble as much as possible real-world processes; *(ii)* at keeping independence from specific representation formalisms, such as Petri nets. The first issue was addressed by using a top-down generation approach (via the definition of a context-free grammar), where well-known workflow patterns are used as building blocks. The probability that a pattern occurs into the generated process model is controlled by the user, via parameters associated to each pattern (i.e., we actually define a stochastic context-free grammar). The second issue is addressed by the generation of a decorated dependency graph as process model. From such graph it is then possible to automatically generate specific formal models, such as a Petri net. Finally, the dependency graph is used to generate traces for the process.

The idea to generate process models for evaluating process mining algorithms is very recent. In [177], van Hee and Zheng, at the Eindhoven University of Technology, present an approach to generate Petri nets representing processes. Specifically, they suggest to use a top-down approach, based on a stepwise refinement of Workflow nets [164], to generate all possible process models belonging to a particular class of Workflow network (Jackson nets). A related approach is presented in [12], where the authors propose to generate Petri nets according to a different set of refinement rules. In both cases, the proposed approach does not address the problem of generating traces from the developed Petri nets. Tools for the simulation of Petri nets, such as CPN Tolls [84, 127], allow to simulate Petri net and generate MXML logs [45]: however, integrating the process generation with their simulation resulted harder than expected.

16.1.1 The Processes Generation Phase

This section presents the procedure for the generation of a business process. In the first subsection we introduce the model used for the description of a generic process and then we present the rules that are involved in the actual generation phase.

The Model for the Process Representation

Since our final aim is the easy generation of models of business processes, we decided to use a very general formalism for our process model description. Petri net models are unambiguous and in-depth studied tools for process modeling, however controlling the generation of a complex process model via refinement of a Petri net may not be so easy for an inexperienced user. For this reason, we decided to model our processes via dependency graphs. A dependency graph can be defined as a graph:

$$G = (V, E, a_{start} \in V, a_{end} \in V)$$

where V is the set of vertices and E is the set of edges. The two vertices a_{start} and a_{end} are used to represent the "start" and the "end" activities of the process model.

Each vertex represents an activity of the process (with all its possible attributes, such as author, duration, ...), while an edge $e \in E$ going from activity a_1 to a_2 represents a dependency relationship between the two activities (i.e. a_2 can start only after that activity a_1 is completed). Let's now define, in a straightforward way, the concept of "incoming activities" and of "exiting activities".

Consider $v \in V$, the set of incoming activities is defined as:

$$\text{in}(v) = \{v_i \mid (v_i, v) \in E\}.$$

Consider $v \in V$, the set of exiting (or outgoing) activities is defined as:

$$\text{out}(v) = \{v_i \mid (v, v_i) \in E\}.$$

From these two definitions we can now define two other simple concepts. Consider $v \in V$, the value of the fan-in for v is defined as: $\overset{\rightarrow}{} \deg(v) = |\text{in}(v)|$, i.e., the number of edges entering in v. Consider $v \in V$, the value of the fan-in for v is defined as: $\deg^{\rightarrow}(v) = |\text{out}(v)|$, i.e. the number of edges exiting from v.

In order to be able to correctly represent a parallel execution of more activities (AND) and a mutual exclusion among the execution of more activities (XOR), we need to define the functions $\mathcal{T}_{out} : V \rightarrow \{AND, XOR\}$ and $\mathcal{T}_{in} : V \rightarrow \{AND, XOR\}$ which have the following meaning. For any vertex (i.e. activity) a with $\deg^{\rightarrow}(a) > 1$, $\mathcal{T}_{out}(a) = AND$ specifies that the flow has to jointly follow all the outgoing edges, while $\mathcal{T}_{out}(a) = XOR$ specifies that the flow has to follow only one of the outgoing edges. Similarly, for any activity a with $^{\rightarrow}\deg(a) > 1$, $\mathcal{T}_{in}(a) = AND$ specifies that the activity has to wait the flow from all the incoming edges before to start, while $\mathcal{T}_{in}(a) = XOR$ specifies that the activity has to wait the flow from just one of the incoming edges before to start.

Using only these two basic types, we can model many real-cases, e.g. a not-exclusive choice among activities $a_1, \ldots, a_n$ can be modeled by an XOR activity, where each outgoing edge leads to one AND activity for each possible proper subset of the activities $a_1, \ldots, a_n$.

Generation of Random Processes

The "decorated" dependency graphs just defined can be used as general representations for the description of relations between activities. In this work, however, we are interested in using them to describe business processes that are assembled by some common and well-known basic "patterns". The basic patterns we consider are the followings (they correspond to the first patterns described in [133]):

- the direct succession of two workflows;
- the execution of more workflows in parallel;
- the mutual exclusion choice between some workflows;
- the repetition of a workflow after another workflow has been executed (as for "preparing" the repetition).

Clearly, these patterns do not describe all the possible behaviors that can be observed in reality, but we think that many realistic processes can be generated from them.

The idea is to start from these simple patterns and to build a complete process in terms of them. We decided to implement this idea via a grammar whose productions are related with the patterns mentioned above. Specifically, we define the context-free grammar $G_{Process} = \{V, \Sigma, R, P\}$, where $V = \{P, G, G', G_{\leftarrow\!\wp}, G_{\wedge}, G_{\otimes}, A\}$ is the set of the not-terminal symbols, $\Sigma = \{; , (,), \leftarrow\!\wp, \wedge, \otimes, a_{start}, a_{end}, a, b, c, \ldots\}$ is the set of all terminals (their "interpretation" is described in Table 16.1), and R is the set of productions:

$$P \rightarrow a_{start} ; \; G \; ; \; a_{end}$$
$$G \rightarrow G' \mid G_{\leftarrow\!\wp}$$
$$G' \rightarrow A \mid (G; G) \mid (A; G_{\wedge}; A) \mid (A; G_{\otimes}; A)$$
$$G_{\leftarrow\!\wp} \rightarrow (G' \leftarrow\!\wp G)$$
$$G_{\wedge} \rightarrow G \wedge G \mid G \wedge G_{\wedge}$$

$$G_\otimes \to G \otimes G \mid G \otimes G_\otimes$$
$$A \to a \mid b \mid c \mid \dots$$

and P is the starting symbol for the grammar.

Using the above-given grammar, a process is described by a string derived from P. It must contain a starting and a finishing activity and, in the middle, a sub-graph G. A sub-graph can be either a "single sub-graph" or a "repetition of a sub-graph". Let's start from the first case, a sub-graph G' can be: a single activity A; the sequential execution of two sub-graphs $(G; G)$; or the execution of some activities in *AND* $(A; G_\wedge; A)$ or *XOR* $(A; G_\otimes; A)$ relation. It is important to note that the generation of parallel and mutual exclusion edges is "well structured", in the sense that there is always a "split activity" and a "join activity" that starts and ends the edges. It should also be mentioned that the system treats the two patterns $(A; G_\wedge; A)$ and $(A; G_\otimes; A)$ in a special way, since it sets the value of $\mathscr{T}_{out}$ of the activity generated by the first occurrence of A, to be equal to the value of $\mathscr{T}_{in}$ of the activity generated by the second occurrence of A, i.e. *AND* for $(A; G_\wedge; A)$ and *XOR* for $(A; G_\otimes; A)$.

The repetition of a sub-graph $(G' \hookleftarrow G)$ is described as follows: each time we want to repeat the "main" sub-graph G', we have to perform another sub-graph G; the idea is that G (that can even be only a single activity) corresponds to the "roll-back" activities required in order to prepare the system to repetition of G'.

The structure of $G_\wedge$ and $G_\otimes$ is simple and expresses the parallel execution or the choice between at least 2 sub-graphs. Finally, A is the set of alphabetic identifiers for the activities (actually, this describes only the generation of the activity name, but the implemented tool "decorates" it with other attributes, such as a unique identifier, the author or the duration). With this definition of the grammar, there could be more activities with the same name, however all the activities are considered to be different.

We provide a complete example where all the steps involved in the generation of a process are shown. The derivation tree, presented in Fig. 16.1, defines the following string of terminals:

Table 16.1 All the terminal symbols of the grammar and their meanings.	Symbols	Meaning
	$()$	used for describing precedence of the operators
	$x; y$	sequential connection of x and y
	$x \wedge y$	parameters executed in parallel (*AND*)
	$x \otimes y$	parameters executed in mutual exclusion (*XOR*)
	$x \hookleftarrow y$	repetition of the first parameter x (the second one, y, can be considered as a "rollback operation")
	a_{start}	"start-up" activity
	a_{end}	"conclusion" activity
	$a, b, c \dots$	activity names

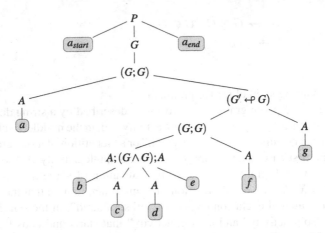

Fig. 16.1 Example of derivation tree. Note that, for space reason, we have omitted the explicit representation of some basic productions.

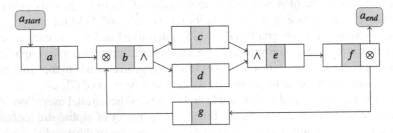

Fig. 16.2 The dependency graph that describes the generated process. Each activity is composed of 3 fields: the middle one contains the name of the activity; the left hand one and the right hand one contain, respectively, the value of the $\mathscr{T}_{in}$ and $\mathscr{T}_{out}$.

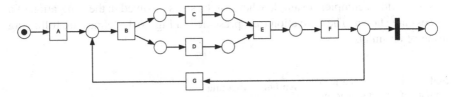

Fig. 16.3 Conversion of the dependency graph of Fig. 16.2 into a Petri net.

$$a_{start}; (a; ((b; (c \wedge d); e; f) \looparrowright g)); a_{end}$$

which can be represented as the dependency graph of Fig. 16.2. Finally, it is possible to convert this process into the Petri net shown in Fig. 16.3.

Grammar Extension with Probabilities

In order to allow the user to control the complexity of the generated processes, we added a probability to each production. This made possible the introduction of user's defined parameters to control the probability of occurrence into the generated process of a specific pattern. In particular, the user has to specify the following probabilities:

$$\pi_1 \text{ for } G' \to A \qquad \pi_3 \text{ for } G' \to (A; G_\wedge; A)$$
$$\pi_2 \text{ for } G' \to (G; G) \qquad \pi_4 \text{ for } G' \to (A; G_\otimes; A)$$
$$\pi_l \text{ for } G_{\leftarrow\rho} \to (G' \leftarrow\rho G)$$

In addition, for both the parallel pattern and the mutual exclusion pattern, our tool requires the user to specify the maximum number of edges ($m_\wedge$ and $m_\otimes$) and the probability distribution that calculates the number of branches to generate. The system will generate, for each AND-XOR split/join, a number of forks between 2 and $m_\wedge$ (or $m_\otimes$, depending on which construct the system is populating) according to the given probability distribution.

At the current stage, the system supports the following probability distributions: (i) uniform distribution; (ii) standard normal (Gaussian) distribution; (iii) beta distribution (with α and β as parameters). These probability distribution generate values between 0 and 1 that are scaled into the correct interval ($2 \ldots m_\wedge$ or $2 \ldots m_\otimes$) and these values indicate the number of branches to generate.

16.1.2 Execution of a Process Model

As already stated, in order to evaluate process mining algorithms, we are not only interested in the generation of a process, but we also need observations of the activities executed at each process instance, i.e. logs. Here we explain how to produce these logs from a generated process. Please, recall that each activity is considered to be different (a unique identifier is associated to it), even if more activities may have the same name. Moreover, in order to facilitate the generation of logs, each time the system chooses the production $G' \to A; (G \wedge G); A$ for the derivation of a process description, it adds to the "split" activity (i.e. the first occurrence of A) a supplementary field with a link to the "join" activity (i.e. the second occurrence of A). Consider, for example, the substring $a; (b \wedge c); d$, with join(a) it is possible to obtain the activity d. The algorithm for the generation uses also a stack, with the standard functions top (checks the first element, without removing it), push (adds a new element into the stack) and pop (removes the first element from the stack).

The procedure used for the generation of an execution of a given process is shown in Algorithm 8. The only operation performed is the call of Algorithm 9 on the first activity of the process using a void stack.

Algorithm 9 is a recursive procedure used to record the execution of the input activity and its successors (via a recursive invocation of the procedure). The two

---

Algorithm 8. ProcessTracer, Execution of a given process.

Input: P: the process model (the dependency graph)

1 $a \leftarrow$ starting activity of P (the a_{start} action)
2 ActivityTracer(a, $\emptyset$)   `/* described in Algorithm 9 */`

Algorithm 9. ActivityTracer, Execution of an activity and all its successors.

Input: a: the current activity; s: stack (LIFO queue) of activities

```
1  if s = ∅ or top(s) ≠ a then
2  │    RecordActivity(a)                        /* described in Algorithm 10 */
3  │    if deg→(a) = 1 then
4  │    │    ActivityTracer(out(a), s)                       /* recursive call */
5  │    else if deg→(a) > 1 then
6  │    │    if 𝒯out(a) = XOR then
7  │    │    │    a1 ← random(out(a))                       /* rnd outgoing act */
8  │    │    │    ActivityTracer(a1, s)                      /* recursive call */
9  │    │    else if 𝒯out(a) = AND then
10 │    │    │    aj ← join(a)                    /* join of the current split */
11 │    │    │    push(s, aj)
12 │    │    │    foreach ai ∈ out(a) do
13 │    │    │    │    ActivityTracer(ai, s)                 /* recursive call */
14 │    │    │    end
15 │    │    │    pop(s)
16 │    │    │    ActivityTracer(aj, s)                      /* recursive call */
17 │    │    end
18 │    end
19 end
```

---

input parameters represent the current activity to be recorded and a stack containing stopping activities (i.e., activities for which the execution of the procedure has to stop), respectively. The last parameter is used when there is an AND split: an instance of the procedure is called for every edge but it must stop when the AND join is reached because, from there on, only one instance of the procedure can continue.

As we said, the procedure of Algorithm 9 has to record the execution of an activity and then it has to call itself on the following activity: if the current activity a is the last one (so $\deg^{\rightarrow}(a) = 0$) then it can terminate; if a is in a sequence (so $\deg^{\rightarrow}(a) = 1$) then it has just to call the same algorithm on the successor activity. Once we reach a split, for example a XOR split (a mutual exclusion), the system has just to choose a random activity and call itself on it. In the last case, the system has to consider the AND split (parallel executions): with this situation it must "execute" all the successor activities (in a not-specific order) but, in order to execute the AND join, all successor activities must be completed. For this purpose, when the procedure is called in the AND split, an extra parameter is passed: this parameter tells the algorithm to stop just before reaching the AND join (actually this parameter is a LIFO (Last In First Out) queue because there can be more AND split/join nested) and then it continues the execution from the join.

In the case of a split activity, the system chooses randomly the activity to follow but all the branches are not equally probable: when the process is generated, each edge, exiting from a split, is augmented with a "weight": the sum of all the weights exiting from the same split is always 1. The random selection considers these weights as probabilities (the higher is the weight, the more probable is the selection of the corresponding branch). As just said, these weight are assigned at the creation of the process in this way: all the exiting branches are (alphabetically) sorted according to the name of the activity they are entering into, and then the weights are assigned according to a given probability distribution (as for the generation of the number of branches, the probability distribution available are: uniform, standard normal and beta). It is necessary to describe the meaning of "selecting a random branch" in the two possible cases (XOR and AND). If the split is a XOR selection, then the meaning is straightforward: if a branch is selected, it is the only one that is allowed to continue the execution; if it is an AND-split, then the procedure will sort all the branches (according to the weight/probability) and later will execute all the activities in the given order.

The cases just described (discriminating on the value of $\deg^{\rightarrow}$) are the only ones the algorithm has to take care of, because all other "high level" patterns are described in terms of them. For example, a loop is expressed as a XOR split (where an edge "continues" to the next activity and the other goes back to the first activity to be performed again). In case of a XOR split, the selection of the branch to be executed is random; so if there is a loop (modelled as a XOR split) the execution is guaranteed to terminate, because the probability of repeating the steps goes to 0.

Algorithm 10 describes the procedure used to record an activity. It uses the extra information of the activity, like its duration (if the activity is not instantaneous), and it decides when an "error" must be inserted (this is required to simulate real-case logs). In this context, an error can be either the swap between the starting time and the end time of an activity or the removal of the activity itself.

Algorithm 10. RecordActivity, Decoration and registration of an activity.

Input: a: the activity to be recorded
1 **if** *activity has to last a time span* **then**
2 | Set the end time for a
3 **end**
4 **if** *activity has to be an "error"* **then**
5 | **if** *the error is a "swap"* **then**
6 | | Swap start time with the end time of a
7 | **else if** *the error is a removal* **then**
8 | | $a \leftarrow$ null
9 | **end**
10 **end**
11 **if** $a \neq$ *null* **then**
12 | Register the execution of a
13 **end**

16.2 Implementation

The entire procedure described in this chapter has been implemented in several tools, developed in Java language. The implementation is formed by two main components: a library (PLGLib) with all the functions currently implemented and a visual tool, for the generation of processes. We find important to have a library that can be easily imported into other projects and that can be used for the batch generation of processes. In order to have a deep control on the generated processes, we added another parameter (with respect to the pattern probabilities): the maximum "depth". Through this, the user can control the maximum number of not-terminals to generate. Suppose the user sets the parameter to the value d; once the grammar has nested d instances of G', then the only not-terminal that can be generated is A. With this parameter there is the possibility to limit the maximum "depth" of the final process.

In the development of this tool, we tried to reuse as many libraries as possible from the ProM tool. For example, we use its sources for the rendering of the networks and for the conversion into a Petri net. For storing the execution logs we use MXML. In the visual interface, we also implemented the calculation of two metrics for the new generated process, described in [15] (Extended Cardoso metric and the Extended cyclomatic one).

In Fig. 16.4 there is a sample Java program that uses PLGLib to generate a random process without specifying all the parameters (so, using the default values) except for the maximum depth parameter. After the generation of the new process, we create a new log, with 10 execution instances for the process and store it into a zipped file. After this operation, the program stores the process into a file with extension ".plg" (this is a zipped file containing an XML representation of the process), in order to allow the loading of the same process for future use. Other functionalities of the library are: the generation of the corresponding Heuristics net (dependency graph), of the corresponding Petri net, the exportation of the process as dot files [52], and the calculation of the metrics cited above. Finally, let us recapitulate the current implementations of PLG:

1. PLG standalone[1]: a software that allows to generate random process models (saving its representation as Heuristics net, Petri net and dependency graph or saving the Petri net in a TPN file) and then can execute such model in order to generate an MXML log file;
2. PLG-CLI[2]: command-line version of the PLG standalone, that is useful for the generation of large datasets;
3. PLG-plugin[3] a plugin for ProM 6.2, with the same functionalities of PLG standalone, but integrated in the current version of ProM.

---

[1] The software can be downloaded for free and with its Java source code from the website http://www.processmining.it/.

[2] The software can be downloaded for free from http://www.processmining.it/.

[3] In the current distribution of ProM 6.2, http://www.promtools.org.

```
1    import it.unipd.math.plg.models.PlgObservation;
2    import it.unipd.math.plg.models.PlgProcess;
3    import java.io.IOException;
4
5
6    class PlgTest {
7      public static void main(String[] args) {
8        try {
9          // define the new process
10         PlgProcess p = new PlgProcess("test process");
11         // randomly populate the new process
12         p.randomize(3);
13         // genearte 10 executions and saves them in a ZIP file
14         p.saveAsNewLog("./test-log.zip", 10);
15         // save the generated process, in order to reuse it
16         p.saveProcessAs("./test-process.plg");
17       } catch (IOException e) {
18         e.printStackTrace();
19       }
20     }
21   }
```

Fig. 16.4 A sample program for the batch generation of business processes using the PLGLib library.

The three version of the PLG are based on the same library "PLGLib", which can be downloaded with PLG standalone. Screenshots of the two versions with a graphical user interface (PLG standalone and PLG-plugin) are presented in Fig. 16.5.

16.3 Summary

In this chapter, we have proposed an approach for the generation of random business processes to ease the evaluation of process mining algorithms. The proposed approach is based on the generation of process descriptions via a (stochastic) context-free grammar whose definition is based on well-known process patterns; each production of this grammar is associated with a probability and the system generates the processes according to these values. In addition, we proposed and commented an algorithm for the generation of execution instances.

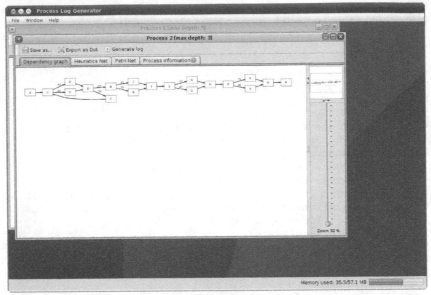

(a) PLG standalone

(b) PLG-plugin for ProM

Fig. 16.5   The software PLG. In (a) there is the standalone version, in (b) there is the ProM plugin.

Part IV
A New Challenge in Process Mining

This part introduces a new challenge for process mining algorithms: online process mining. Here we are going to analyze some control-flow mining algorithms which are capable of mining control-flows from event streams.

Chapter 17
Process Mining for Stream Data Sources

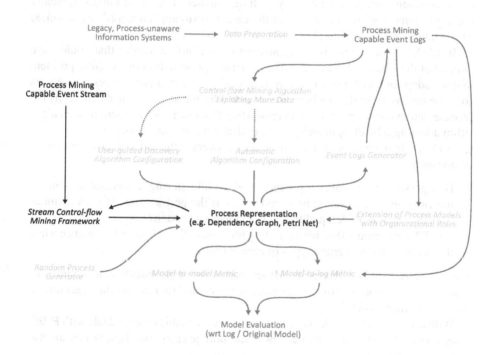

The number of installations of Information Systems is increasing more and more. These systems produce huge amounts of log data. In some cases it is impossible to store all event data in a classical log file, simply for storage limitations. Moreover, while only few processes are in steady-state and, most of the processes, due to changing circumstances, evolve and systems need to be flexible.

However most process discovery algorithms have been defined for processing complete log files of a single unchanging process. As a result, existing algorithms are difficult to apply in such evolving environments. In this chapter, we will discuss peculiarities of mining streaming event data in the context of process mining.

© Springer International Publishing Switzerland 2015
A. Burattin: *Process Mining Techniques in Business Environments*, LNBIP 207,
DOI 10.1007/978-3-319-17482-2_17

In particular, we present algorithms for discovering process models based on streaming event data [27, 28]. In the rest of this chapter we refer to this problem as *Streaming Process Discovery* (or SPD).

According to [2, 14], a data stream consists of an unbounded sequence of data items with a very high throughput. In addition to that, the following assumptions are typically made: *(i)* data is assumed to have a small and fixed number of attributes; *(ii)* mining algorithms should be able to process an infinite amount of data, without exceeding memory limits or otherwise fail, no matter how many items are processed; *(iii)* for classification tasks, data has a limited number of possible class labels; *(iv)* the amount of memory available to a learning/mining algorithm is considered finite, and typically much smaller than the data observed in a reasonable span of time; *(v)* there is a small upper bound on the time within the algorithm is allowed to process an item, e.g. algorithms have to scale linearly with the number of processed items: typically the algorithms work with one pass of the data; *(vi)* stream "concepts" are assumed to be stationary or evolving [178, 184].

In SPD, a typical task is to reconstruct a control-flow model that could have generated the observed event log. The general representation of the SPD problem that we adopt in this context is shown in Fig. 17.1: one or more sources emit events (represented as solid dots) which are observed by the stream miner that keeps the representation of the process model up-to-date. Obviously, no standard mining algorithm adopting a batch approach is able to deal with this scenario.

An SPD algorithm has to give satisfactory answers to the following two categories of questions:

1. Is it possible to discover a process model while storing a minimal amount of information? What should be stored? What is the performance of such methods both in terms of model quality and speed/memory usage?
2. Can SPD techniques deal with evolving processes? What is the performance when the stream exhibits certain types of concept drift?

We will show how the Heuristics Miner, one of the more effective algorithms for practical applications of process mining, can be adapted for stream mining according to our SPD framework.

With respect to the problems mentioned in Sect. 1.2, this chapter deals with **P-05**: issues connected to computational power and storage capacity. These issues are also discussed in Sect. 8.5.

17.1 Basic Concepts

The main difference between classical process mining [170] and SPD lies in the assumed input format. For SPD we can assume that streaming event data may even come from multiple sources rather than a static event log containing historic data.

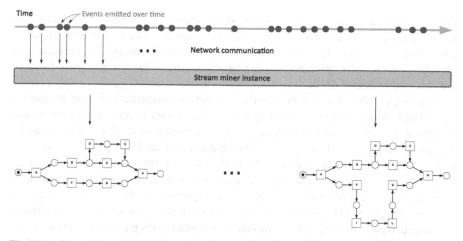

Fig. 17.1 General idea of SPD: the stream miner continuously receives events and, using the latest observations, updates the process model.

In this context, we assume that each *event*, received by the miner, contains the *name of the activity* executed, the *case id* it belongs to, and a *timestamp*. A formal definition of these elements is:

Definition 17.1 (Activity, Case, Time and Event Stream). Let $\mathscr{A}$ be a set of activities and $\mathscr{C}$ be a set of case identifiers. An *event* is a triplet $(c, a, t) \in \mathscr{C} \times \mathscr{A} \times \mathbb{N}$, i.e., the occurrence of activity a for case c (i.e. the process instance) at time t (timestamp of emission of the event). Actually, in the miner, rather than using an absolute timestamp, we consider a progressive number representing the number of events seen so far, so an event at time t is followed by another event at time $t + 1$, regardless the time lasts between them. $S \in (\mathscr{C} \times \mathscr{A} \times \mathbb{N})^*$ is an event stream, i.e., a sequence of events that are observed item by item. The events in S are sorted according to the order they are emitted, i.e. the event timestamp.

Starting from this definition, it is possible to define some functions:

Definition 17.2 (Case time scope). $t_{start}(c) = \min_{(c,a,t) \in S} t$, i.e. the time when the first activity for c is observed. $t_{end}(c) = \max_{(c,a,t) \in S} t$, i.e. the time when the last activity for c is observed.

Definition 17.3 (Subsequence). Given a sequence of events $S \in (\mathscr{C} \times \mathscr{A} \times \mathbb{N})^*$, it is a sorted series of events: $S = \langle \dots, s_i, \dots, s_{i+j}, \dots \rangle$ where $s_i = (c, a, t) \in \mathscr{C} \times \mathscr{A} \times \mathbb{N}$. A subsequence S_i^j of S is a sequence that identifies the elements of S starting at position i and finishing at position $i + j$: $S_i^j = \langle s_i, \dots, s_{i+j} \rangle$.

In order to compare classical control-flow discovery algorithms with new algorithms for streams, we can consider an *observation period*. An observation period O for an event stream S, is a finite subsequence of S starting at time i and with size j:

$O = S_i^j$. Basically, any observation period is a finite subsequence of a stream, and it can be understood as a classical log file (although the "head" and "tail" of some cases may be missing). A well-established control-flow discovery algorithm that can be applied to an observation period log is the Heuristics Miner, whose main features are reported in Sect. 5.1.

In analogy with classical data streams, an event stream can be defined as *stationary* or *evolving*. In our context, a stationary stream can be seen as generated by a business process that does not change with time. On the contrary, an evolving stream can be understood as generated by a process that changes in time. More precisely, different modes of change can be considered: *(i)* drift of the process model; *(ii)* shift of the process model; *(iii)* cases (i.e., execution instances of the process) distribution change. Drift and shift of the process model correspond to the classical two modes of *concept drift* [17] in data streams: a drift of the model refers to a gradual change of the underlying process, while a model shift happens when a change between two process models is more abrupt. The change in cases distribution represents another way in which an event stream can evolve, i.e. the original process may stay the same during time, however, the distribution of the cases is not stationary. By this we mean that the distribution of the features of the process cases change with time. For example, in a production process of a company selling clothing, the items involved in incoming orders (i.e., cases features) during winter will follow a completely different distribution with respect to items involved in incoming orders during the summer. Such distribution change may significantly affect the relevance of specific paths in the control-flow of the involved process.

Going back to process model drift, there is a peculiarity of business event streams that cannot be found in traditional data streams. An event log records that a specific activity a_i of a business process P has been executed at time t for a *specific* case c_j. If the drift from P to P' happens at time t^* *while* the process is running, there might be cases for which all the activities have been executed within P (i.e., cases that have terminated their execution before t^*), cases for which all the activities have been executed within P' (i.e., cases that have started their execution on or after t^*), and cases that have some activities executed within P and some others within P' (i.e., cases that have started their execution before t^* and have terminated after t^*). We will refer to these cases as *transient cases*. So, under this scenario, the stream will first emit events of cases executed within P, followed by events of transient cases, followed by events of cases executed within P'. On the contrary, if the drift *does not occur* while the process is running, the stream will first report events referring to complete executions (i.e. cases) of P, followed by events referring to complete executions of P' (no transient cases). In any case, the drift is characterized by the fact that P' is very similar to P, i.e. the change in the process which emits the events is limited.

Due to space limitation, we restrict our treatment to stationary streams and streams with concept drift with no generation of transient cases. The treatment of other scenarios is left for future work.

17.2 Heuristics Miners for Streams

In this section, we are presenting variants of the Heuristics Miner algorithm to address the SPD problem under different scenarios. First of all, we present two basic algorithms where the standard batch version of Heuristics Miner is used on logs, which are built as "observation periods" extracted from the stream. These algorithms will be used as a baseline reference for the experimental evaluation. Subsequently, a "fully online" version of Heuristics Miner, to cope with stationary streams, drift of the process model with no transient cases, and shift of the process model, is introduced.

17.2.1 Baseline Algorithm for Stream Mining

The simplest way to adapt the Heuristics Miner algorithm to deal with streams is to collect events during specific observation periods and then applying the batch version of the algorithm to the current log. This idea is described by Algorithm 11 in which two different policies to maintain events in memory are considered. Specifically, an event e from the stream S is observed ($e \leftarrow observe(S)$) and analyzed ($analyze(e)$) to decide if the event has to be considered for mining. If this is the case, it is checked whether there is room in memory to accommodate the event. If the memory is full ($size(M) = max_M$) then the memory policy given as input is adopted. Two different policies are considered: *periodic resets*, and *sliding windows* [2, Chap. 8]. In the case of *periodic resets*, all the events contained in memory are deleted (*reset*), while in the case of *sliding windows*, only the oldest event is deleted (*shift*). Subsequently, e is

---

Algorithm 11. Sliding Window HM / Periodic Resets HM

Input: S event stream; M memory of size max_M; P_M memory policy (can be '*reset*' or '*shift*')

```
1  forever do
2  |   e ← observe(S)              /* observe an event, where e = (c_i, a_i, t_i) */
   |   /* Check if event e has to be used                                        */
3  |   if analyze(e) then
   |   |   /* Memory update                                                      */
4  |   |   if size(M) = max_M then
5  |   |   |   if P_M is reset then reset(M)
6  |   |   |   if P_M is shift then shift(M)
7  |   |   end
8  |   |   insert(M, e)
   |   |   /* Mining update                                                      */
9  |   |   if perform mining then
10 |   |   |   HeuristicsMiner(M)
11 |   |   end
12 |   end
13 end
```

---

inserted in memory and it is checked if it is necessary to perform a mining action. If
mining has to be performed, the Heuristics Miner algorithm is executed on the events
in memory ($HeuristicsMiner(M)$). Graphical representations of the two policies are
reported in Fig. 17.2.

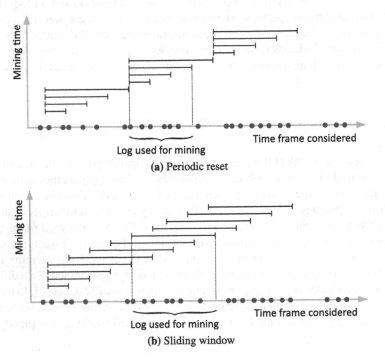

Fig. 17.2 Two basic approaches for the definition of a finite log out of a stream of events. The
horizontal segments represent the time frames considered for the mining.

A potential advantage of the two policies described consists in the possibility to
mine the log not only by using Heuristics Miner, but also using any process mining
algorithm already available for traditional batch process discovery techniques (not
only for control-flow discovery, for example it is possible to extract information
about the social network). However, the notion of "history" is not very accurate:
only the more recent events are considered, and an equal importance is assigned to
all of them. Moreover, the model is not updated in real-time, since each new event
received triggers only the update of the log, not necessarily also an update of the
model: performing a model update for each new event would result in a significant
computational burden, well outside the computational limitations assumed for a
true online approach. In addition to that, the time required by these approaches is
completely unbalanced: when a new event arrives, only inexpensive operations are
performed; instead, when the model needs to be updated, the log retained in memory
is mined from scratch. So, every event is handled at least twice: the first time to store

it into a log and subsequently any time the mining phase takes place on it. In an online setting, it is more desirable a procedure that does not need to process each event more than once ("one pass algorithm" [138]).

17.2.2 Stream-Specific Approaches

In this section, we suggest how to modify the scheme of the basic approaches, so to implement a real online framework, the final approach is described in Algorithm 12. In this framework: the "current" log is described in terms of "latest observed activities" and "latest observed dependencies". Specifically, we define three queues:

1. $Q_{\mathscr{A}}$, with entries in $\mathscr{A} \times \mathbb{R}$, stores the most recent observed activities jointly with a weight for each activity (that represents its degree of importance with respect to mining);
2. $Q_{\mathscr{C}}$, with entries in $\mathscr{C} \times \mathscr{A}$, stores the most recent observed event for each case;
3. $Q_{\mathscr{R}}$ with entries in $\mathscr{A} \times \mathscr{A} \times \mathbb{R}$, stores the most recent observed direct succession relations jointly with a weight for each succession relation (that represents its degree of importance with respect to mining).

These queues are used by the online algorithm to retain the information needed to perform mining.

The detailed description of the new algorithm is presented in Algorithm 12. Specifically, the algorithm runs forever, considering, at each round, the current observed event $e = (c_i, a_i, t_i)$. For each current event, it is checked if a_i is already in $Q_{\mathscr{A}}$. If this is not the case, a_i is inserted in $Q_{\mathscr{A}}$ with weight 0. If a_i is already present in the queue, it is removed from its current position and moved at the beginning of the queue. In any case, before insertion, it is checked if $Q_{\mathscr{A}}$ is full. If this is the case, the oldest stored activity, i.e. the last in the queue, is removed. Subsequently, the weights of $Q_{\mathscr{A}}$ are updated by f_{W_A}. After that, queue $Q_{\mathscr{C}}$ is examined to look for the most recent event observed for case c_i. If a pair (c_i, a) is found, it is removed from the queue, an instance of the succession relation (a, a_i) is created and searched in $Q_{\mathscr{R}}$. If it is found, it is moved from the current position to the beginning of $Q_{\mathscr{R}}$. If it is a new succession relation, its weight is set to 0. In any case, before insertion, it is checked if $Q_{\mathscr{R}}$ is full. If this is the case, the oldest stored relation, i.e. the last in the queue, is removed. Subsequently, the weights of $Q_{\mathscr{R}}$ are updated by f_{W_R}. Next, after checking if $Q_{\mathscr{C}}$ is full (in which case the oldest stored event is removed), the event e is stored in $Q_{\mathscr{C}}$.

Finally, it is checked if a model has to be generated. If this is the case, the procedure $generateModel(Q_{\mathscr{A}}, Q_{\mathscr{R}})$ is executed taking as input the current version of queues $Q_{\mathscr{A}}$ and $Q_{\mathscr{R}}$ and producing "classical" model representations, such as Causal Nets [149] or Petri nets.

Algorithm 12 is parametric with respect to: $i)$ in which way weights of queues $Q_{\mathscr{A}}$ and $Q_{\mathscr{R}}$ are updated by $f_{W_{\mathscr{A}}}$, $f_{W_{\mathscr{R}}}$, respectively; $ii)$ how a model is generated by $generateModel(Q_{\mathscr{A}}, Q_{\mathscr{R}})$. In the following, $generateModel(\cdot, \cdot)$ will correspond to the procedure defined by Heuristics Miner. In particular, it is possible to consider

---

Algorithm 12. Online HM

Input: S event stream; $max_{Q_{\mathscr{A}}}$, $max_{Q_{\mathscr{C}}}$, $max_{Q_{\mathscr{R}}}$ maximum memory sizes for queues $Q_{\mathscr{A}}$,
$Q_{\mathscr{C}}$, and $Q_{\mathscr{R}}$, respectively; $fw_{\mathscr{A}}$, $fw_{\mathscr{R}}$ model policy; $generateModel(\cdot, \cdot)$.

```
 1  forever do
 2  │   e ← observe(S)          /* observe a new event, where e = (cᵢ, aᵢ, tᵢ) */
    │                           /* check if event e has to be used            */
 3  │   if analyze(e) then
 4  │   │   if ∄(a, w) ∈ Q_𝒜 s.t. a = aᵢ then
 5  │   │   │   if size(Q_𝒜) = max_Q_𝒜 then
 6  │   │   │   │   removeLast(Q_𝒜)              /* removes last entry of Q_𝒜 */
 7  │   │   │   end
 8  │   │   │   w ← 0
 9  │   │   else
10  │   │   │   w ← get(Q_𝒜, aᵢ)  /* get returns the old weight w of aᵢ and
    │   │   │   removes (aᵢ, w) */
11  │   │   end
12  │   │   insert(Q_𝒜, (aᵢ, w))                /* inserts in front of Q_𝒜 */
13  │   │   Q_𝒜 ← fw_𝒜(Q_𝒜)                     /* updates the weights of Q_𝒜 */
14  │   │   if ∃(c, a) ∈ Q_𝒞 s.t. c = cᵢ then
15  │   │   │   a ← get(Q_𝒞, cᵢ)     /* get returns the old activity a of cᵢ
    │   │   │   and removes (cᵢ, a) */
16  │   │   │   if ∄(aₛ, a_f, u) ∈ Q_𝓡 s.t. (aₛ = a) ∧ (a_f = aᵢ) then
17  │   │   │   │   if size(Q_𝓡) = max_Q_𝓡 then
18  │   │   │   │   │   removeLast(Q_𝓡)          /* removes last entry of Q_𝓡 */
19  │   │   │   │   end
20  │   │   │   │   u ← 0
21  │   │   │   else
22  │   │   │   │   u ← get(Q_𝓡, a, aᵢ)         /* get returns the old weight u of
    │   │   │   │   relation a → aᵢ and removes (a, aᵢ, u) */
23  │   │   │   end
24  │   │   │   insert(Q_𝓡, (a, aᵢ, u))         /* inserts in front of Q_𝓡 */
25  │   │   │   Q_𝓡 ← fw_𝓡(Q_𝓡)                 /* updates the weights of Q_𝓡 */
26  │   │   else if size(Q_𝒞) = max_Q_𝒞 then
27  │   │   │   removeLast(Q_𝒞)                  /* removes last entry of Q_𝒞 */
28  │   │   end
29  │   │   insert(Q_𝒞, (cᵢ, aᵢ))               /* inserts in front of Q_𝒞 */
    │   │   /* generate model                                                  */
30  │   │   if model then
31  │   │   │   generateModel(Q_𝒜, Q_𝓡)
32  │   │   end
33  │   end
34  end
```

$Q_{\mathscr{A}}$ as the counter of activities (to filter out only the most frequent ones) and $Q_{\mathscr{R}}$ as the counter of direct succession relations, which are used for the computation of the dependency values between pairs of activities. The following subsections present some specific instances for $fw_{\mathscr{A}}$ and $fw_{\mathscr{R}}$.

Online Heuristics Miner (Stationary Streams)

In the case of stationary streams, we can reproduce the behavior of Heuristics Miner as follows. $Q_{\mathscr{A}}$ should contain, for each activity a, the number of occurrences of a observed in S till the current time. Similarly, $Q_{\mathscr{R}}$ should contain, for each succession (a, b), the number of occurrences of (a, b) observed in S till the current time. Thus both $fw_{\mathscr{A}}$ and $fw_{\mathscr{R}}$ must just increment the weight of the first element of the queue:

$$fw_{\mathscr{A}}((a, w)) = \begin{cases} (a, w + 1) & \text{if } first(Q_{\mathscr{A}}) = (a, w) \\ (a, w) & \text{otherwise} \end{cases}$$

$$fw_{\mathscr{R}}((a, b, w)) = \begin{cases} (a, b, w + 1) & \text{if } first(Q_{\mathscr{R}}) = (a, b, w) \\ (a, b, w) & \text{otherwise} \end{cases}$$

where $first(\cdot)$ returns the first element of the queue.

In case of stationary streams, it is possible to use the Hoeffding bound to derive error bounds on the measures computed by the online version of Heuristics Miner. These bounds became tighter and tighter with the increase of the number of processed events. Section 17.3 reports some details on that.

It must be noticed that, if the sizes of the queues are large enough, the Online Heuristics Miner collects all the needed statistics from the beginning of the stream till the current time. So it performs very well, considering that the activity distribution of the stream is stationary. However, in real world business processes, it is natural to observe variations both in events distribution and in the workflow of the process generating the stream (concept drift).

In order to cope with concept drift, more importance should be given to more recent events than to older ones. In the following we present a variant of Online Heuristics Miner able to do that.

Online Heuristics Miner with Aging (Evolving Streams)

The idea, in this case, is to decrease the weights for the events (and relations) over time when they are not observed. So, every time a new event is observed, only the weight of its activity (and observed succession) is increased, all the others are reduced. Given an "aging factor" $\alpha \in [0, 1)$, the weight functions $fw_{\mathscr{A}}$ (for activities) and $fw_{\mathscr{R}}$ (for succession relations) are modified so to replace all the occurrences of w on the right side of the equality with αw:

$$fw_{\mathscr{A}}((a, w)) = \begin{cases} (a, (\alpha w) + 1) & \text{if } first(Q_{\mathscr{A}}) = (a, w) \\ (a, \alpha w) & \text{otherwise} \end{cases}$$

$$fw_{\mathscr{R}}((a, b, w)) = \begin{cases} (a, b, (\alpha w) + 1) & \text{if } first(Q_{\mathscr{R}}) = (a, b, w) \\ (a, b, \alpha w) & \text{otherwise} \end{cases}$$

The basic idea of these new functions is to decrease the "history" (i.e., the current number of observations) by an aging factor α (in the formula: αw) before increasing it by 1 (the new observation).

These new functions decrease all the weights associated to either an event or a succession relation according to the aging factor α, which determines the "speed" in forgetting an activity or succession relation, however the most recent observation (the first in the respective queue) is increased by 1. Notice that, if an activity or succession relation is not observed for t time steps, its weight becomes α^t. Thus the value of α allows to control the speed of "forgetting": the closer α is to 0, the faster the weight associated to an activity (or succession relation) that has not been observed for some time goes to 0, thus to allow the miner to assign larger values to recent events. In this way the miner is more sensitive to sharp variations of the event distribution (concept shift); however the output (generated models) may be less stable because the algorithm becomes more sensitive to random fluctuations of the sampling distribution. When the value of α is close to 1, activities that have not been observed recently, but were seen more often some time ago, are able to retain their significance, thus allowing the miner to be able to cope with mild variations of the event distribution (concept drift), but not so reactive in case of concept shift.

One drawback of this approach is that, while it is able to "forget" old events, it is not able, at time t, to preserve precise statistics for the last k observations and to completely drop observations occurred before time $t - k$. This ability could be useful in case a sudden drastic change in the event distribution.

Online Heuristics Miner with Self-adapting Aging (Evolving Stream)

The third approach explored in this section introduces α as a parameter to control the importance of the "history" for the mining: the closer it is to 1, the more importance is given to the history. The value of α, should be decided according to the known degree of "not-stationarity" of the stream; however, this information might not be available or it might not be fixed (for example, the process is stationary for a period, then it evolves, and then it becomes stationary again). To handle these cases, it is possible to dynamically adapt the value of α. In particular, the idea is *to lower the value of α when a drift is observed and to increase it when the stream seems to be stationary*.

A possible approach to detect the drift is to monitor for variations on the fitness value. This measure, evaluated at a certain period, can be considered as the amount of events (considering only the latest ones) that the current mined process is able to explain. When the fitness value changes drastically, it is likely that a drift has occurred. Using the drift detection, it is possible to adapt α according to the following rules:

- if the fitness *decreases* (i.e. there is a drift) α should decreases too (up to 0), in order to allow the current model to adapt to the new data;
- if the fitness *remains unchanged* (i.e. it is within a small interval), it means that there is no drift so the value of α should be increased (up to 1);
- if the fitness *increases*, α should be increased too (up to 1).

The experiments, presented on the next section, consider only variations of α by a constant factor. Alternative update policies (e.g. making the speed of change of α proportional to the observed fitness change) are not shown here, but can be considered fur future investigations.

17.2.3 Stream Process Mining with Lossy Counting (Evolving Stream)

The approach presented in this section is an adaptation of an existing technique, used to approximate frequency count. In particular, we modified the "Lossy Counting" algorithm described in [102]. We preferred this approach to Sticky Sampling (described in the same paper) since authors stated that, in practice, Lossy Counting performs better. The entire procedure is presented in Algorithm 13.

The basic idea of Lossy Counting algorithm is to conceptually divide the stream into buckets of width $w = \lceil \frac{1}{\varepsilon} \rceil$, where $\varepsilon \in (0, 1)$ is an error parameter. The *current* bucket (i.e., the bucket of the last element seen) is identified with $b_{current} = \lceil \frac{N}{w} \rceil$, where N is the progressive events counter.

The basic data structure used by Lossy Counting is a set of entries of the form (e, f, Δ) where: e is an element of the stream; f is the estimated frequency of the item e; and Δ is the maximum possible error. Every time a new element e is observed, the algorithm looks whether the data structure contains an entry for the corresponding element. If such entry exists, then its frequency value f is incremented by 1, otherwise a new tuple is added: $(e, 1, b_{current} - 1)$. Every time $N \equiv 0 \mod w$, the algorithm cleans the data structure by removing the entries that satisfy the following inequality: $f + \Delta \le b_{current}$. Such condition ensures that, every time the cleanup procedure is executed, $b_{current} \le \varepsilon N$.

This algorithm has been adapted to the SPD problem, using three instances of the basic data structure. In particular, it counts the frequencies of the activities (with the data structure $\mathcal{D}_A$) and the frequencies of the direct succession relations (with the data structure $\mathcal{D}_R$). In order to obtain the relations, a third instance of the same data structure is used, $\mathcal{D}_C$. In $\mathcal{D}_C$, each item is of the type (c, a, f, Δ) where $c \in \mathcal{C}$ represents the case identifier; f and Δ, as in previous cases, respectively correspond to the frequency and to the bucket id; and $a \in A$ is the latest activity observed on the corresponding case. Every time a new activity is observed, $\mathcal{D}_A$ is updated. After that, the procedure checks if, given the case identifiers of the current event, there is an entry in $\mathcal{D}_C$. If this is not the case a new entry is added to $\mathcal{D}_C$ (by adding the current case id and the activity observed). Otherwise, the f and a components of the entry in $\mathcal{D}_C$ are updated.

Algorithm 13. Lossy Counting HM

Input: S event stream; N the bucket counter (initially value 1); $\mathcal{D}_A$ activities set; $\mathcal{D}_C$ cases set; $\mathcal{D}_R$ relations set; $generateModel(\cdot, \cdot)$.

1   $w \leftarrow \lceil \frac{1}{\varepsilon} \rceil$ /* define the bucket width */

2   **forever do**

3      $b_{current} = \lceil \frac{N}{w} \rceil$ /* define the current bucket id */

4      $e \leftarrow observe(S)$ /* observe a new event, where $e = (c_i, a_i, \Delta_i)$ */

       /* update the $\mathcal{D}_A$ data structure   */

5   **if** $\exists (a, f, \Delta) \in \mathcal{D}_A$ *such that* $a = a_i$ **then**

6   Remove the entry (a, f, Δ) from $\mathcal{D}_A$

7        $\mathcal{D}_A \leftarrow (a, f + 1, \Delta)$ /* updates the frequency of element a_i */

8   **else**

9        $\mathcal{D}_A \leftarrow \mathcal{D}_A \cup \{(a_i, 1, b_{current} - 1)\}$ /* inserts the new observation */

10   **end**

       /* update the $\mathcal{D}_C$ data structure   */

11   **if** $\exists (c, a, f, \Delta) \in \mathcal{D}_C$ *such that* $c = c_i$ **then**

12   Remove the entry (c, a, f, Δ) from $\mathcal{D}_C$

13       $\mathcal{D}_C \leftarrow (c, a_i, f + 1, \Delta)$ /* updates the frequency and last activity of case c_i */

       /* update the $\mathcal{D}_R$ data structure   */

14   Build relation r_i as $a \rightarrow a_i$

15   **if** $\exists (r, f, \Delta) \in \mathcal{D}_R$ *such that* $r = r_i$ **then**

16   Remove the entry (r, f, Δ) from $\mathcal{D}_R$

17         $\mathcal{D}_R \leftarrow (r, f + 1, \Delta)$ /* updates the frequency of element r_i */

18   **else**

19         $\mathcal{D}_R \leftarrow \mathcal{D}_R \cup \{(r_i, 1, b_{current} - 1)\}$ /* adds the new observation */

20   **end**

21   **else**

22       $\mathcal{D}_C \leftarrow \mathcal{D}_C \cup \{(c_i, a_i, 1, b_{current} - 1)\}$ /* adds the new observation */

23   **end**

       /* periodic cleanup   */

24   **if** $N = 0 \mod w$ **then**

25   **foreach** $(a, f, \Delta) \in \mathcal{D}_A$ *such that* $f + \Delta \leq b_{current}$ **do**

26   Remove (a, f, Δ) from $\mathcal{D}_A$

27   **end**

28   **foreach** $(c, a, f, \Delta) \in \mathcal{D}_C$ *such that* $f + \Delta \leq b_{current}$ **do**

29   Remove (c, a, f, Δ) from $\mathcal{D}_C$

30   **end**

31   **foreach** $(r, f, \Delta) \in \mathcal{D}_R$ *such that* $f + \Delta \leq b_{current}$ **do**

32   Remove (r, f, Δ) from $\mathcal{D}_R$

33   **end**

34   **end**

35     $N \leftarrow N + 1$ /* increments the bucket counter */

       /* generate model   */

36   **if** *model* **then**

37       $generateModel(\mathcal{D}_A, \mathcal{D}_R)$

38   **end**

39 **end**

The Heuristics Miner can be used to generate the model, since a set of dependencies between activities is available.

17.3 Error Bounds on Online Heuristics Miner

If we assume a stationary stream, i.e. a stream where the distribution of events does not change with time (no concept drift), then it is possible to give error bounds on the measures computed by the online version of Heuristics Miner.

In fact, let's consider an execution of the online Heuristics Miner on the stream S. Let $Q_{\mathscr{A}}(t)$, $Q_{\mathscr{C}}(t)$, and $Q_{\mathscr{R}}(t)$ be the content of the queues used by Algorithm 12 at time t. Let $case_{overlap}(t) = \{c \in \mathscr{C} \mid t_{start}(c) \leq t \wedge t_{end}(c) \geq t\}$ be the set of cases that are *active* at time t; $\Delta_c = \max_t |case_{overlap}(t)|$; $n_c(t)$ be the cumulative number of cases which have been removed from $Q_{\mathscr{C}}(t)$ during the time interval $[0, t]$; and $nc(t) = |Q_{\mathscr{C}}(t)| + n_c(t)$. Given two activities a and b, let $\rho_{ab} \in [0, \xi_{ab}]$ be the random variable reporting the number of successions (a, b) contained in a randomly selected trace in S. With $\mathscr{A}_S$ and $\mathscr{R}_S$ we denote the set of activities and successions, respectively, observed within the entire stream S. Then it is possible to state the following theorem:

Theorem 17.1 (Error bounds). *Let* $(a \Rightarrow_S b)$, $a \Rightarrow_S (b \wedge c)$, *be the measures computed by the Heuristics Miner algorithm on a time-stationary stream S, and* $(a \Rightarrow_{S_0^t} b)$, $a \Rightarrow_{S_0^t} (b \wedge c)$, *be the measures computed at time t by the online version of the Heuristics Miner algorithm on the stream S. If* $\max_A \geq |A_S|$, $\max_R \geq |R_S|$, $\max_C \geq \Delta_c$, *then with probability* $1 - \delta$ *the following bounds hold:*

$$(a \Rightarrow_S b)\left(\frac{E[\rho_{ab} + \rho_{ba}]}{E[\rho_{ub} + \rho_{ba}] + \varepsilon_{ab}(t) + \frac{1}{nc(t)}}\right) -$$

$$\frac{\varepsilon_{ab}(t)}{E[\rho_{ab} + \rho_{ba}] + \varepsilon_{ab}(t) + \frac{1}{nc(t)}} \leq (a \Rightarrow_{S_0^t} b)$$

$$(a \Rightarrow_{S_0^t} b) \leq (a \Rightarrow_S b)\left(\frac{E[\rho_{ab} + \rho_{ba}]}{E[\rho_{ab} + \rho_{ba}] - \varepsilon_{ab}(t) + \frac{1}{nc(t)}}\right) +$$

$$\frac{\varepsilon_{ab}(t)}{E[\rho_{ab} + \rho_{ba}] - \varepsilon_{ab}(t) + \frac{1}{nc(t)}}$$

And, similarly, for $a \Rightarrow (b \wedge c)$:

$$(a \Rightarrow_S (b \wedge c)) \left(\frac{E[\rho_{bc} + \rho_{cb}]}{E[\rho_{ab} + \rho_{ac}] + \varepsilon_{abc}(t) + \frac{1}{nc(t)}} \right) -$$

$$\frac{\varepsilon_{bc}(t)}{E[\rho_{ab} + \rho_{ac}] + \varepsilon_{abc}(t) + \frac{1}{nc(t)}} \leq (a \Rightarrow_{S_0^t} (b \wedge c))$$

$$(a \Rightarrow_{S_0^t} (b \wedge c)) \leq (a \Rightarrow_S (b \wedge c)) \left(\frac{E[\rho_{bc} + \rho_{cb}]}{E[\rho_{bc} + \rho_{cb}] - \varepsilon_{abc}(t) + \frac{1}{nc(t)}} \right) +$$

$$\frac{\varepsilon_{bc}(t)}{E[\rho_{ab} + \rho_{ac}] - \varepsilon_{abc}(t) + \frac{1}{nc(t)}}$$

where $\forall d, e, f \in A_S$, $\varepsilon_{de}(t) = \sqrt{\frac{(\xi_{de} + \xi_{ed})^2 \ln(2/\delta)}{2nc(t)}}$, $\varepsilon_{def}(t) = \sqrt{\frac{(\xi_{de} + \xi_{df})^2 \ln(2/\delta)}{2nc(t)}}$, and $E[x]$ is the expected value of x.

Proof. Let consider the Heuristics Miner definition $(a \Rightarrow_S b) = \frac{|a>_S b| - |b>_S a|}{|a>_S b| + |b>_S a| + 1}$ (as presented in Eq. 11.1). Let N_c be the number of cases contained in S_0^t, then

$$(a \Rightarrow_{S_0^t} b) = \frac{|a >_{S_0^t} b| - |b >_{S_0^t} a|}{|a >_{S_0^t} b| + |b >_{S_0^t} a| + 1} = \frac{\frac{|a>_{S_0^t} b| - |b>_{S_0^t} a|}{N_c}}{\frac{|a>_{S_0^t} b| + |b>_{S_0^t} a|}{N_c} + \frac{1}{N_c}}$$

and

$$(a \Rightarrow_S b) = \lim_{N_c \to +\infty} \frac{\frac{|a>_{S_0^t} b| - |b>_{S_0^t} a|}{N_c}}{\frac{|a>_{S_0^t} b| + |b>_{S_0^t} a|}{N_c} + \frac{1}{N_c}} = \frac{E[\rho_{ab} - \rho_{ba}]}{E[\rho_{ab} + \rho_{ba}]}.$$

We recall that $\overline{X} = \frac{|a>_{S_0^t} b| - |b>_{S_0^t} a|}{N_c}$ is the mean of the random variable $X = (\rho_{ab} - \rho_{ba})$ computed over N_c independent observations, i.e. traces, and that $X \in [-\xi_{ba}, \xi_{ab}]$. We can then use the *Hoeffding* bound [81] that states that, with probability $1 - \delta$

$$\left| \overline{X} - E[X] \right| < \varepsilon_X = \sqrt{\frac{r_X^2 \ln\left(\frac{2}{\delta}\right)}{2N_c}},$$

where r_X is the range of X, which in our case is $r_X = (\xi_{ab} + \xi_{ba})$.

By using the *Hoeffding* bound also for the variable $Y = (\rho_{ab} + \rho_{ba})$, we can state that with probability $1 - \delta$

$$\frac{E[X] - \varepsilon_X}{E[Y] + \varepsilon_Y + \frac{1}{N_c}} \leq \frac{\overline{X}}{\overline{Y} + \frac{1}{N_c}} = (a \Rightarrow_{S_0^t} b),$$

which after some algebra can be rewritten as

$$\frac{E[X]}{E[Y]}\left(\frac{E[Y]}{E[Y]+\varepsilon_Y+\frac{1}{N_c}}\right)-\frac{\varepsilon_X}{E[Y]+\varepsilon_Y+\frac{1}{N_c}}\le(a\Rightarrow_{s_0^t}b)$$

By observing that $(a\Rightarrow_s b)=\frac{E[X]}{E[Y]}$, $r_X=r_Y=(\xi_{ab}+\xi_{ba})$, and that at time t, under the theorem hypotheses, no information is removed from the queues and $N_c=nc(t)$, the first bound is proved. The second bound can be proved starting from

$$(a\Rightarrow_{s_0^t}b)\le\frac{E[X]+\varepsilon_X}{E[Y]-\varepsilon_Y+\frac{1}{N_c}}.$$

The last two bounds can be proved in a similar way by considering $X=(\rho_{bc}+\rho_{cb})\in[0,\xi_{bc}+\xi_{cb}]$ and $Y=(\rho_{ab}+\rho_{ac})\in[0,\xi_{ab}+\xi_{ac}]$, which leads to $\varepsilon_X=\sqrt{\frac{(\xi_{bc}+\xi_{cb})^2\ln(2/\delta)}{2N_c}}$ and $\varepsilon_Y=\sqrt{\frac{(\xi_{ab}+\xi_{ac})^2\ln(2/\delta)}{2N_c}}$.

Similar bounds can be obtained also for the other measures computed by Heuristics Miner. From the bounds it is possible to see that, with the increase of the number of observed cases $nc(t)$, both $\frac{1}{nc(t)}$ and the errors $\varepsilon_{ab}(t)$ and $\varepsilon_{abc}(t)$ go to 0 and the measures computed by the online version of Heuristics Miner consistently converge to the "right" values.

17.4 Results

The algorithms presented in this chapter have been tested using four datasets: event logs from two artificial processes (one stationary and one evolving); a synthetic example; and a real event log.

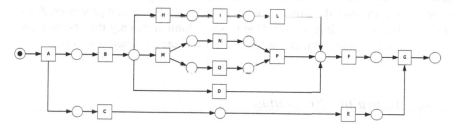

Fig. 17.3 Model 1. Process model used to generate the stationary stream.

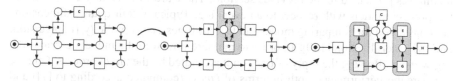

Fig. 17.4 Model 2. The three process models that generate the evolving stream. Red rounded rectangles indicate areas subject to modification (Color figure online).

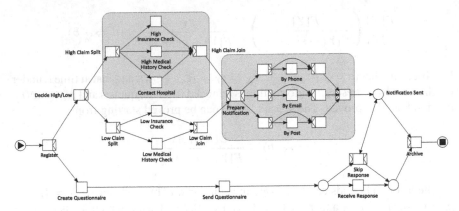

Fig. 17.5 Model 3. The first variant of the third model. Red rounded rectangles indicate areas that will be subject to the modifications (Color figure online).

17.4.1 Models Description

The two artificial processes are shown in Figs. 17.3 and 17.4, both are described in terms of a Petri net. The first one (Model 1) describes the complete model that is simulated to generate the stationary stream. The second one (Model 2) presents the three models which are used to generate three logs describing an evolving stream. In this case, the final stream is generated considering the hard shift of the three logs generated from the single process executions.

The synthetic example (Model 3) is reported in Fig. 17.5. This example is taken from [16, Chap. 5] and is expressed as a YAWL [159] process. This model describes a possible health insurance claim process of a travel agency. This example is modified 4 times so, at the end, the stream contains traces from 5 different processes. Also in this case the type of drift is shift. Due to space limitation, only the first process is presented and the red rectangles indicate areas that are modified over time.

17.4.2 Algorithms Evaluation

The streams generated from the described models are used for the evaluation of the techniques presented in the previous sections. There are various metrics to evaluate the process models with respect to an event log. Typically four quality dimensions are considered for comparing model and log: *(a)* fitness, *(b)* simplicity, *(c)* precision, and *(d)* generalization [150]. In order to measure how well the model describes the log without allowing the reply of traces not generated by the target process, here we measure the performance both in terms of *fitness* (computed according to [1]) and in terms of *precision* (computed according to [110]). The first measure reaches its maximum when all the traces in the log are properly replied by the model, while

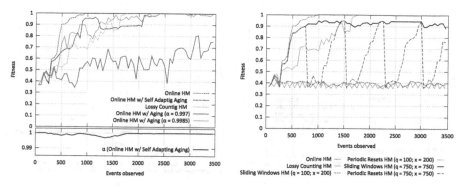

Fig. 17.6 Aggregated experimental results for five streams generated by Model 1. Evolution in time of average fitness for Online HM with queues size 100 and log size for fitness 200; curves for HM with Aging ($\alpha = 0.9985$ and $\alpha = 0.997$), HM with Self Adapting (evolution of the α value is shown at the bottom), Lossy Counting and different configurations of the basic approaches are reported as well.

the second one prefers models that describe a "minimal behavior" with respect to all the models that can be generated starting from the same log. In all experiments, the *fitness* and *precision* measures are computed over the last x observed events (where x varies according to log size), q refers to the maximum size of queues, and default parameters of Heuristics Miner, for model generation, are used.

The main characteristics of the three streams are:

- *Streams for Model 1*: 3448 events, describing 400 cases;
- *Streams for Model 2*: 4875 events, describing 750 cases (250 cases and 2000 events for the first process model, 250 cases and 1750 events for the second, and 250 cases with 1125 events for the third one);
- *Stream for Model 3*: 58783 events, describing 6000 cases (1199 cases and 11838 events for the first variant; 1243 cases and 11690 events for the second variant; 1176 cases and 12157 events for the third variant; 1183 cases and 10473 events for the fourth variant; and 1199 cases and 12625 events for the fifth variant).

We compared the basic approaches versus the different online versions of stream miner, against the different streams.

Model 1

Figure 17.6 reports the aggregated experimental results for the five streams generated by Model 1. The figure presents, on the left hand side, a comparison of the evolution of the average fitness of the Online HM, the HM with Aging ($\alpha = 0.9985$ and $\alpha = 0.997$), the HM with Self Adapting approach and Lossy Counting. For these runs a queue size of 100 has been used and, for the fitness computation, the latest 200 events are considered. In this case, the lossy counting considers an error value $\varepsilon = 0.01$. The right hand side of Fig. 17.6 compares the basic approaches, with different window and

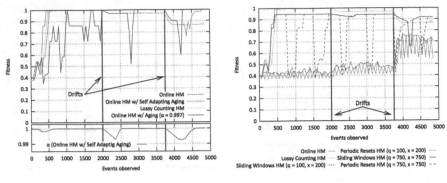

Fig. 17.7 Aggregated experimental results for five streams generated by evolving Model 2. Evolution in time of average fitness for Online HM with queues size 100 and log size for fitness 200; curves for HM with Aging ($\alpha = 0.997$), HM with Self Adapting (evolution of the α value is shown at the bottom), Lossy Counting and different configurations of the basic approaches are reported as well. Drift occurrences are marked with vertical bars.

fitness sizes against the Online HM and the Lossy Counting approach. As expected, since there is no drift, the Online HM outperforms the versions with aging. In fact, HM with Aging, beside being less stable, also degrades performances as the value of α decreases, i.e. less importance is given to less recent events. This is consistent with the bad performance reported for the basic approaches which can exploit only the most recent events contained in the window. The self adapting strategy, after an initial variation of the α parameter, is able to converge to the Online HM by eventually choosing a value of α equals to 1.

Model 2

Figure 17.7 reports the aggregated experimental results for the five streams generated by Model 2. In this case, we adopted exactly the same experimental setup, procedure and results presentation described before. In addition, the occurrences of drift are marked. As expected, the performance of Online HM decreases at each drift, while HM with Aging is able to recover from the drifts. The price paid for this ability is a less stable behavior. HM with Self Adapting aging seems to be the right compromise, eventually able to recover from the drifts while showing a stable behavior. The α curve shows that the self adapting strategy seems to be able to detect the concept drifts.

Model 3

The Model 3, with the synthetic example, has been tested with the basic approaches (Sliding Windows and Periodic Resets), the Online HM, the HM with Self Adapting

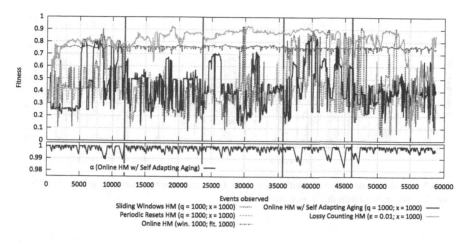

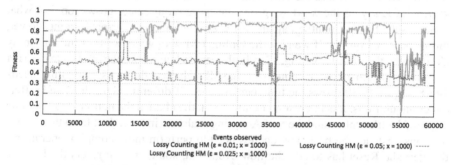

Fig. 17.8 Detailed results of the basic approaches, Online HM, HM with Self Adapting and Lossy Counting (with different configurations) on data of Model 3. Vertical gray lines indicate points where concept drift occur.

and the Lossy Counting and the results are presented in Fig. 17.8. In this case, the Lossy Counting and the Online HM outperform the other approaches. Lossy Counting reaches higher fitness values, however Online HM is more stable and seems to better tolerate the drifts. The basic approaches and the HM with Self Adapting, on the other hand, are very unstable; moreover it is interesting to note that the value of α, of the HM with Self Adapting, is always close to 1. This indicates that the short stabilities of the fitness values are sufficient to increase α, so the updating policy (i.e. the increment/decrement speed of α) presented, for this particular case, seems to be too fast. The second graph, on the bottom, presents three runs of the Lossy Counting, with different values for ε. As expected, the lower the value of the accepted error, the better the performances.

Due to the size of this dataset, it is interesting to evaluate the performance of the approaches also in terms of space and time requirements.

Figure 17.9 presents the average memory required by the miner during the processing of the entire log. Different configurations are tested, both for the basic approaches

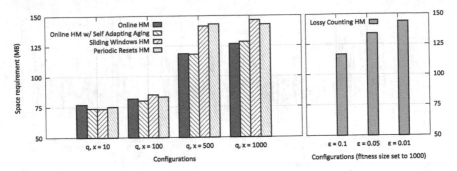

Fig. 17.9 Average memory requirements, in MB, for a complete run over the entire log of Model 3, of the approaches (with different configurations).

with the Online HM and the HM with Self Adapting, and the Lossy Counting algorithm. Clearly, as the windows grow, the space requirement grows too. For what concerns the Lossy Counting, again, as the ε value (accepted error) becomes lower, more space is required. If we pick the Online HM with window 1000 and the Lossy Counting with ε 0.01 (from Fig. 17.8, both seem to behave similarly) the Online HM consumes less memory: it requires 128.3 MB whereas the Lossy Counting needs 143.8. Figure 17.10 shows the time performance of different algorithms and different configurations. It is interesting to note, from the chart at the bottom, that the time required by the Online and the Self Adapting is almost independent from the configurations. Instead, the basic approaches need to perform more complex operations: the Periodic Reset has to add the new event and, sometimes, it resets the log; the Sliding Window has to update the log every time a new event is observed.

In order to study the dependence of the storage requirements of Lossy Counting with respect to the error parameter ε, we have run experiments on the same log for different values of ε, recording the maximum size of the Lossy Counting sets during execution. Results for $x = 1000$ are reported in Fig. 17.11. Specifically, the figure compares the maximum size of the generated sets, the average *fitness* value and the average *precision* value. As expected, as the value of ε becomes larger, both the *fitness* value and the sets size quickly decrease. The *precision* value, on the contrary, initially decreases and then goes up to very high values. This indicates an over-specialization of the model to specific behaviors.

As an additional test, we decide to compare the proposed algorithms under extreme storage conditions which do allow only to retain limited information about the observed events. Specifically, Table 17.1 reports the average time required to process a single event, average *fitness* and *precision* values when queues with size 10 and 100, respectively, are used. For Lossy Counting we have used an ε value which approximately requires sets of similar sizes. Please note that, for this log, a single process trace is longer than 10 events so, with a queue of 10 elements it is not possible to keep in queue all the events of a case (because events of different cases are interleaved). From the results it is clear that, under these conditions, the order of

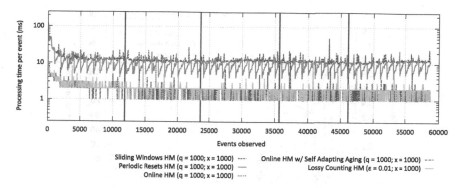

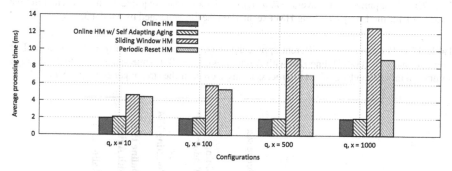

Fig. 17.10 Time performances over the entire log of Model 3. *Top:* time required to process a single event by different algorithms (logarithmic scale). Vertical gray lines indicate points where concept drift occur. *Bottom:* average time required to process an event over the entire log, with different configurations of the algorithms.

occurrence of the algorithms in the table (column order) is inversely proportional to all the evaluation criteria (i.e. execution time, *fitness*, *precision*).

The online approaches presented in this work have been tested also against a real dataset and results are presented in Fig. 17.12. The reported results refer to 9000 events generated from the document management system, by Siav S.p.A., and run on an Italian bank institute. The observed process contains 8 activities and is assumed to be stationary. The mining is performed using a queues size of 100 and, for the fitness computation, the latest 200 events are considered. The behavior of the fitness curves seems to indicate that some minor drifts occur.

As stated before, the main difference between Online HM and Lossy Counting is that, whereas the main parameter of Online HM is the size of the queues (i.e. the maximum space the application is allowed to use), the ε parameter of Lossy Counting cannot control the memory occupancy of the approach. Figure 17.13 proposes two comparisons of the approaches with different configurations, against the real stream dataset. In particular we defined the two configurations so that the average memory required by Lossy Counting and Online HM are very close. The results presented are actually the average values over four runs of the approaches. Please note that

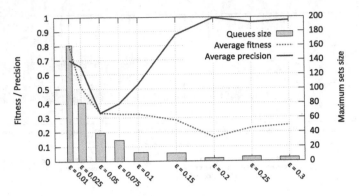

Fig. 17.11 Comparison of the average *fitness*, *precision* and space required, with respect to different values of ε for the Lossy Counting HM executed on the log generated by Model 3.

Table 17.1 Performance of different approaches with queues/sets size of $q = 10$ and $q = 100$ elements and $x = 1000$. Online HM with Aging uses $\alpha^{1/q} = 0.9$. Time values refer to the average number of milliseconds required to process a single event of the stream generated by Model 3.

	Queue size	Sliding Window HM	Lossy Counting HM	Online HM with Aging	Online HM
Average Time (ms)		4.66	2.61	2.11	1.97
Average Fitness	$q = 10$	0.32	0.28	0.32	0.32
Average Precision		0.44	0.87	0.38	0.38
Average Time (ms)		5.79	2.85	1.99	1.91
Average Fitness	$q = 100$	0.32	0.51	0.42	0.74
Average Precision		0.42	0.65	0.68	0.71

the two configurations validates the fitness against different window sizes (in the first case it contains 200 events, in the second one 1000) and this causes the second configuration to validate results against a larger history.

The top part of the figure presents a configuration that uses, on average, about 100 MB. To obtain these performances, several tests have been made and, at the end, for Lossy Counting these parameters have been used: $\varepsilon : 0.2$, fitness queue size: 200. For Online HM, the same fitness is used, but the queue size is set to 500. As the plot shows, it is interesting to note that, in terms of fitness, this configuration is absolutely enough for the Online HM approach instead, for Lossy Counting, it is not. The second plot, at the bottom, presents a different configuration that uses about 170 MB. In this case, the error (i.e. ε) for Lossy Counting is set to 0.01, the queue size of Online HM is set to 1500 and, for both, the fitness queue size is set to 1000. In this case the two approaches generate really close results, in terms of fitness.

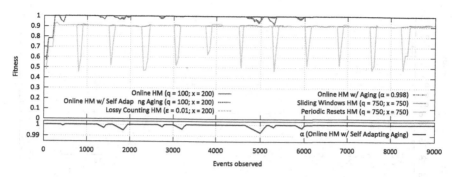

Fig. 17.12 Fitness performance on the real stream dataset by different algorithms.

As final consideration, this empirical evaluation clearly shows that –at least in our real dataset– both Online HM and Lossy Counting are able to reach very high performances, however the Online is able to better exploit the information available with respect to the Lossy Counting. In particular, Online HM considers only a finite number of possible observations (depending on the queue size) that, in this particular case, are sufficient to mine the correct model. The Lossy Counting, on the contrary, keeps all the information for a certain time-frame (obtained starting from the error parameter) without considering how many different behaviors are already seen.

Note on Fitness Measure

The usage of fitness for the evaluation of stream process mining algorithms seems to be an effective choice. However, this might not always be true: let's consider two very different processes P' and P'' and a stream composed of events generated by alternate executions of P' and P''. Under specific conditions, the stream miner will generate a model that contains both P' and P'', connected by an initial XOR-split and merged with a XOR-join. This model will show a very high fitness value (it can replay traces from both P' and P''), however the mined model is not the one expected, i.e. the alteration in time of P' and P'' is not reflected well.

In order to deal with the problem above-mentioned, we propose to perform some approaches also in terms of "precision". This measure is thought to prefer models that describe a "minimal behavior" with respect to all the model that can be generated starting from the same log. In particular, we used the approach by Muñoz-Gama and Carmona described in [110]. Figure 17.14 presents the precision calculated for four approaches during the analysis of the dataset of real events. It should not surprise that the stream specific approaches reach very good precision values, whereas the basic approach with periodic reset needs to recompute, every 1000 events, the model from scratch. It is interesting to note that both Online HM and Lossy Counting are not

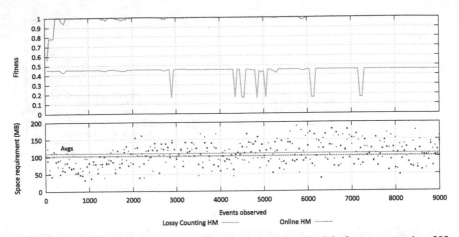

(a) Configuration that requires about 100MB. Lossy Counting: $\varepsilon : 0.2$, fitness queue size: 200; Online HM: queue size: 500, fitness queue size: 200.

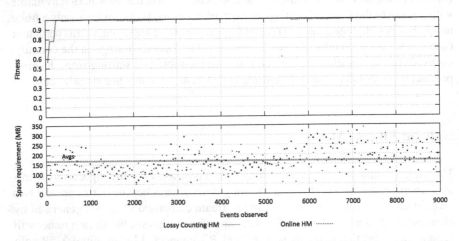

(b) Configuration that requires about 170MB. Lossy Counting: $\varepsilon : 0.01$, fitness queue size: 1000; Online HM: queue size: 1500, fitness queue size: 1000.

Fig. 17.13 Performances comparison between Online HM and Lossy Counting, in terms of fitness and memory consumption.

able to reach the top values, whereas the Self adapting one, after some time, reaches the best precision, even if its value fluctuates a little. The basic approach with sliding window, instead, seems to behave quite nicely, even if the stream specific approaches outperform it.

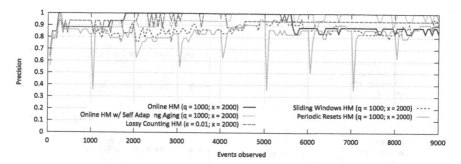

Fig. 17.14 Precision performance on the real stream dataset by different algorithms.

```
1    <log openxes.version="1.0RC7" xes.features="nested-attributes"
         xes.version="1.0" xmlns="http://www.xes-standard.org/">
2        <trace>
3            <string key="concept:name" value="case_id_0" />
4            <event>
5                <date key="time:timestamp"
                     value="2015-01-23T10:33:04.004+02:00" />
6                <string key="concept:name" value="A" />
7                <string key="lifecycle:transition" value="Task_Execution"
8                    />
9            </event>
10       </trace>
11   </log>
```

Listing 17.1 OpenXES fragment streamed over the network.

17.5 Implementation

All the approaches presented in this chapter have been implemented in the ProM
6.1 toolkit. Moreover, a "stream simulator" and a "logs merger" have also been
implemented to allow for experimentation (to test new algorithms and to compose
logs).

Communications between stream sources and stream miner are performed over
the network: each event emitted consists of a "small log" (i.e., a trace which contains
exactly one event), encoded as a XES string. An example of an event log streamed
is presented in Listing 17.1. This approach is useful to simulate "many-to-many
environments" where one source emits events to many miners and one miner can use
many stream sources. The current implementation supports only the first scenario
(currently it is not possible to mine streams generated by more than one source).

Figure 17.15 proposes the set of ProM plugins implemented, and how they interact each other. The available plugins can be split into two groups: plugins for the
simulation of the stream and plugins to mine streaming event data. To simulate a
stream there is the "Log Streamer" plugin. This plugin receives a static log file as
input and streams each event over the network, according to its timestamp (in this
context, timestamps are used only to determine the order of events). It is possible
to define the time between each event, in order to test the miner under different

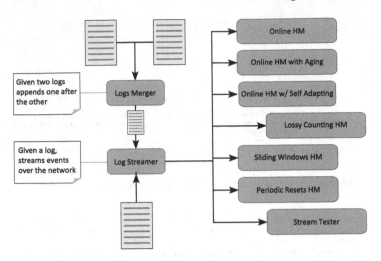

Fig. 17.15 Architecture of the plugins implemented in ProM and how they interact with each other. Each rounded box represents a ProM plugin.

emission rates (i.e. to simulate different traffic conditions). A second plugin, called "Logs Merger" can be used to concatenate different log files generated by different process models, just for testing purposes.

Once the stream is active (i.e. events are sent through the network), the clients can use these data to mine the model. There is a "Stream Tester" plugin, which just shows the events received. The other plugins support the two basic approaches (Subsect. 17.2.1), and the four stream specific approaches (Subsects. 17.2.2 and 17.2.3).

In a typical testing session of a new stream process mining algorithm, we expect to have two separate ProM instances active at the same time: the first streaming events over the network and the second collecting and mining them.

Figure 17.16 contains four screenshots of the ProM plugins implemented. The image on top right, in particular, contains the process streamer: the left bar describes the stream configuration options (such as the speed or the network port for new connections), the central part contains a representation of the log as a dotted chart (the x axis represents the time, and each point with the same timestamp x value is an event occurred at the same instant). Blue dots are the events that are not yet sent (future events), green ones are the events already streamed (past events). It is possible to change the color of the future events so that every event referring to the same activity or to the same process instance has the same color. The figure at bottom left contains the Stream Tester: each event of a stream is appended to this list, which shows the timestamp of the activity, its name and its case id. The left bar contains some basic statistics (i.e. beginning of the streaming session, number of events observed and average number of events observed per second). The last picture, at bottom right, represents the Online HM miner. This view can be divided into three parts: the central part, where the process representation is shown (in this case, as a Causal Net); the left bar contains, on top, buttons to start/stop the miner

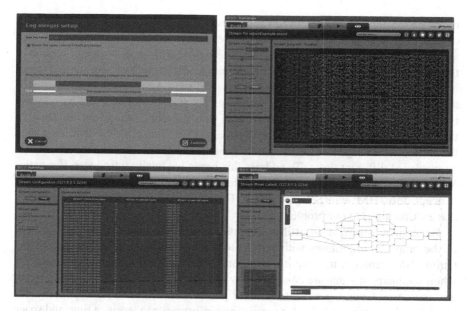

Fig. 17.16 Screenshots of four implemented ProM plugins. The first image (top left) shows the logs merger (it is possible to define the overlap level of the two logs); the second image (top right) represents the log streamer, the bottom left image is the stream tester and the image at the bottom right shows the Online HM.

plus some basic statistics (i.e., beginning of the streaming session, number of events observed and average number of events observed per second); at the bottom, there is a graph which shows the evolution of the fitness measure.

Moreover, Command-Line Interface (CLI) versions of the miners are available too[1]. In these version, events are read from a static file (one event per line) and the miners update the model (this implementation realizes an incremental approach of the algorithm). These implementations can be run in batch and are used for automated experimentation.

17.6 Summary

In this chapter, we faced the problem of discovering processes for streaming event data having different characteristics, i.e. stationary streams and streams with drift.

First, we considered basic window-based approaches, where the standard Heuristics Miner algorithm is applied to statics logs obtained by using a *moving window* on the stream (we considered two different policies). Then, we introduced a framework for stream process mining which allows the definition of different approaches, all

---

[1]See http://www.processmining.it for more details.

based on the dependencies between activities. These can be seen as online versions of the Heuristics Miner algorithm and differentiate from each other in the way they assign importance to the observed events. The Online HM, an incremental version of the Heuristics Miner, gives the same importance to all the observed events, and thus it is specifically apt to mine stationary streams. HM with Aging gives less importance to older events. This is obtained by weighting the statistics of an event by a factor, the α value, which exponentially decreases with the age of the event. Because of that, this algorithm is able to cope with streams exhibiting concept drift. However, the choice of the "right" value for α is difficult, and different values for α could also be needed at different times. To address this issue, we finally introduced Heuristics Miner able to automatically adapt the aging factor on the basis of the detection of concept drift (HM with Self Adapting). Finally, we adapted a standard approach (Lossy Counting) to our problem.

Experimental results on artificial, synthetic and real data showed the efficacy of the proposed algorithms with respect to the basic approaches. Specifically, the Online HM turns out to be quite stable and performs well for streams, especially when stationary streams are considered; while HM with Self Adapting aging factor and the Lossy Counting seem to be the right choice in case of concept drift. The largest log has been used also for measuring performance in terms of time and space requirements.

This work has also been used as fundamental basis for further online algorithms, such as those reported in [21, 99], which are able to perform online discovery of Declare process models from event streams.

With respect to the problems mentioned in Sect. 1.2, this chapter deals with problem **P-05**: issues connected to computational power and storage capacity (see also Sect. 8.5). The results reported in this chapter represent only the first step on this field, and involve only control-flow discovery. Additional work has to be done in order to improve the quality of actual mining algorithms, and to allow conformance analysis and enhancement applicable on event streams.

Part V
Conclusions and Future Work

Chapter 18
Conclusions and Future Work

In this book, we identified several key problems that emerge when trying to deploy a process mining project in an organization. The entire work can be considered as a case study, in which we had the opportunity to closely work with some local companies. This situation gave us the opportunity to continuously improve our comprehension of the actual situation.

18.1 Wrap-Up

We pointed out several problems that might emerge during different stages of a project. Specifically, before starting the actual process mining phase, all the information must be available. Typically this is happening, so it is necessary to construct such data. Once the log is ready, it is possible to configure mining algorithms parameters. However, our experience with companies demonstrates that this is a challenging task: users, that are non-expert in process mining, find many difficulties in driving the mining through the parameters of the algorithm. Finally, when a model is extracted, it is necessary to evaluate the result, with respect to the ideal model (e.g., protocols or guidelines) or with respect to the log used for mining. Moreover, a process model might be more useful if information on activities originators and business roles is added too. Apart from all these problems, there is one concerning small and medium enterprises: these companies might have difficulties in storing all the events generated by their information systems and the analysis of such data can really deal with high computational power consumes. These problems were analyzed in detail in Chaps. 8 and 9.

Figure 18.1 shows all contributions (red italic font) described in this book with their corresponding input and output dependencies (black font). These contributions are numbered (gray font, in parenthesis) so we can reference them in the following subsections.

© Springer International Publishing Switzerland 2015
A. Burattin: *Process Mining Techniques in Business Environments*, LNBIP 207,
DOI 10.1007/978-3-319-17482-2_18

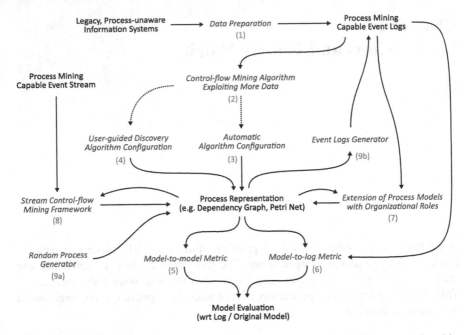

Fig. 18.1 Contributions, written in red italic font, presented in this book. They are numbered in order to be referenced in the text. Dotted lines indicate that the input/output is not an "object" for the final user, instead it represents a methodological approach (e.g., a way to configure parameters) (Color figure online).

Problems with Data Preparation

The first point we addressed with data preparation, is the lack of the case ID field in the log. In this book we presented a solution, formalized using a relational algebra approach, to extract this information from decorative meta-data fields *(Contribution (1) in Fig. 18.1, reported in Chap. 10).*

Problems at Mining Time

We defined a generalization of one of the most used process mining control-flow algorithms (namely, Heuristics Miner). In our new version, the algorithm is able to deal with activities recorded as time intervals, in order to improve the quality of mined models *(Contribution (2) in Fig. 18.1, reported in Chap. 11).* Concerning the configuration of parameters, we considered the set of parameters of our new mining algorithm (which is the same set of the Heuristics Miner) and we proposed a procedure for the "conservative discretization" of the possible values of such parameters ("conservative" in the sense that surely we do not lose any possible output model). This discrete set of possible values induces a finite number of mined models.

In order to select the "best" model, out of this set, we proposed two approaches: one completely autonomous *(Contribution (3) in Fig.* 18.1, *reported in Chap.* 12*)* and another which requires user's interaction *(Contribution (4) in Fig.* 18.1, *reported in Chap.* 13*)*. In the latter case, processes are clustered in a hierarchy and the analyst is required to navigate the hierarchy by iteratively selecting between two branches until a satisfying result is obtained.

Problems with Results Evaluation

Concerning the mining results evaluation, we proposed two new metrics: one model-to-model metric, and another which is model-to-log. In the first case *(Contribution (5) in Fig.* 18.1, *reported in Sect.* 15.1*)*, we improved an already available model-to-model metric, in order to make it more suitable for comparing models generated by process mining algorithms. We applied this new metric to the procedure for the exploration of process hierarchies. The model-to-log metric *(Contribution (6) in Fig.* 18.1, *reported in Sect.* 15.2*)*, instead, allows the analyst to compare a Declare model with respect to a given log. Healthiness measures are also proposed in order to have numerical values of adherence of the entire model (or of single constraints) to the available observations (i.e., the log).

Extension of Process Models with Business Roles

In this book we have analyzed an approach for the extension of business process models with roles *(Contribution (7) in Fig.* 18.1, *reported in Chap.* 14*)*. Specifically, our approach, given a process model and a log, tries to find handover of roles in order to partition activities in "swim-lanes". Handover of roles are discovered using specific metrics and two thresholds allow the system to be robust against presence of noise in the log.

Store Capacity and Computational Power Requirements

To address issues related to store capacity and computation power requirements, we proposed an online solution which allows the incremental mining of event streams *(Contribution (8) in Fig.* 18.1, *reported in Chap.* 17*)*. In particular, our approach provides a framework which is able to incrementally mine a stream of events. Different instances of this framework are proposed and compared too. All these solutions have been characterized by their ability to deal with concept drifts: some of them perform better when concept drifts occur, other are only apt to deal with stationary stream.

Lack of Data

In this book we also took into account problems related with the lack of experimental data. We proposed a system which is able to generate random business processes *(Contribution (9a) in Fig.* 18.1, *reported in Subsect.* 16.1.1*)* and can simulate them

(Contribution (9b) in Fig. 18.1, *reported in Subsect.* 16.1.2*)*, producing log files. This data can be used for evaluation of new process mining algorithms.

18.2 Future Work

It is possible to continue the work reported in this book in several directions. Specifically, we expect that, in real scenarios, several other problems could emerge and therefore require attention and effort.

Concerning the identification of the case ID, we think it would be interesting to consider not only the *value* of the case ID candidates but, to go deeper, their *semantic meaning* (if any), which could act as *a-priori* knowledge. Moreover, a flexible framework for expressing and feeding the system with *a-priori* knowledge is desirable, in order to earn a higher level of generalization. Then, other refinements are domain-specific: dealing with documents, for instance, we could exploit their content in order to confirm or reject the findings of our algorithms, when the result confidence is low.

The automatic extraction of the best model can be improved by increasing the number of explored hypothesis. In particular, a dynamic generation of the hypothesis space could help to cope with the corresponding computational burden. Another improvement that we think could be very useful, is the introduction of machine learning techniques, to allow the system to "acquire" the preferable process patterns in the search, and to improve the "goodness measure" by directly encoding this information.

A lot of work is still required on the stream framework; in particular, we are willing to conduct a deeper analysis on the influence of the different parameters on the presented approaches. Moreover, it would be interesting to extend the current approach to mining the organizational perspective of the process. Finally, from a process analyst point of view, it may be interesting not only to show the current updated process model, but also to report the "evolution points" of the process.

Finally, we think that also the process and log generator can be extensively improved: concerning generation of processes, a characterization of the space of the processes generated by our approach could be useful, so that who is going to use the system, may exactly know which space of processes it generates. Another open issue is how much the generated processes can be considered "realistic": since using process patterns increases the probability to generate a realistic process, it would be nice to have control on this factor. An idea, for tackling this problem, could be to establish the values of some complexity measures for a given dataset of real business processes and to constrain the generation of random processes to these values for the given metrics. Concerning the execution of a process, we think that an important improvement is the generation of decorative attributes (such as originator or activities-related data), in order to simulate a more realistic execution.

References

1. Adriansyah, A., van Dongen, B., van der Aalst, W.M.P.: Conformance checking using cost-based fitness analysis. In: 2011 IEEE 15th International Enterprise Distributed Object Computing Conference, pp. 55–64. IEEE, August 2011. (Cited on page 192)
2. Aggarwal, C.: Data Streams: Models and Algorithms. Advances in Database Systems, vol. 31. Springer, Boston (2007). (Cited on pages 54, 178, and 181)
3. Agrawal, R., Gunopulos, D., Leymann, F.: Mining process models from workflow logs. In: Schek, H.-J., Saltor, F., Ramos, I., Alonso, G. (eds.) EDBT 1998. LNCS, vol. 1377, pp. 469–483. Springer, Heidelberg (1998). (Cited on pages 12, 36, and 38)
4. Agrawal, R., Srikant, R.: Fast algorithms for mining association rules. In: International Conference on Very Large Data (1994). (Cited on page 29)
5. Aiolli, F., Burattin, A., Sperduti, A.: A business process metric based on the alpha algorithm relations. In: Daniel, F., Barkaoui, K., Dustdar, S. (eds.) BPM Workshops 2011, Part I. LNBIP, vol. 99, pp. 141–146. Springer, Heidelberg (2012). (Cited on page 139)
6. Allen, J.F.: Maintaining knowledge about temporal intervals. Commun. ACM **26**(11), 832–843 (1983). (Cited on page 90)
7. Bae, J., Liu, L., Caverlee, J., Zhang, L.-J., Bae, H.: Development of distance measures for process mining, discovery, and integration. Int. J. Web Serv. Res. **4**(4), 1–17 (2007). (Cited on pages 52 and 138)
8. Barricelli, N.A.: Esempi numerici di processi di evoluzione. Methodos **6**, 45–68 (1954). (Cited on page 43)
9. Bauer, A., Leucker, M., Schallhart, C.: Runtime verification for LTL and TLTL. ACM Trans. Softw. Eng. Methodol. **20**(4), 1–64 (2011). (Cited on page 147)
10. Beer, I., Ben-david, S., Eisner, C., Rodeh, Y.: Efficient detection of vacuity in temporal model checking. Formal Methods Syst. Des. **18**(2), 141–163 (2001). (Cited on page 148)
11. Bergenthum, R., Desel, J., Lorenz, R., Mauser, S.: Process mining based on regions of languages. In: Alonso, G., Dadam, P., Rosemann, M. (eds.) BPM 2007. LNCS, vol. 4714, pp. 375–383. Springer, Heidelberg (2007). (Cited on page 45)
12. Bergmann, G., Horváth, Á., Ráth, I., Varró, D.: A benchmark evaluation of incremental pattern matching in graph transformation. In: Ehrig, H., Heckel, R., Rozenberg, G., Taentzer, G. (eds.) ICGT 2008. LNCS, vol. 5214, pp. 396–410. Springer, Heidelberg (2008). (Cited on page 164)
13. Berry, M.J.A., Linoff, G.S.: Data Mining Techniques, 2nd edn. Wiley Computer Publishing, New York (2004). (Cited on page 27)
14. Bifet, A., Holmes, G., Kirkby, R., Pfahringer, B.: MOA: massive online analysis learning examples. J. Mach. Learn. Res. **11**, 1601–1604 (2010). (Cited on page 178)

15. Lassen, K.B., van der Aalst, W.M.P.: Complexity metrics for workflow nets. Inf. Softw. Technol. **51**(3), 610–626 (2009). (Cited on page 172)
16. Chandra Bose, R.P.J.: Process Mining in the Large: Preprocessing, Discovery, and Diagnostics. Ph.d. thesis, Technische Universiteit Eindhoven (2012). (Cited on page 192)
17. Bose, R.P.J.C., van der Aalst, W.M.P., Žliobaitė, I., Pechenizkiy, M.: Handling concept drift in process mining. In: Mouratidis, H., Rolland, C. (eds.) CAiSE 2011. LNCS, vol. 6741, pp. 391–405. Springer, Heidelberg (2011). (Cited on pages 55 and 180)
18. Brualdi, R.A.: Introductory Combinatorics, 5th edn. Pearson Prentice Hall, Old Tappan (2009). (Cited on pages 122 and 128)
19. Buijs, J.: Flexible Evolutionary Algorithms for Mining Structured Process Models. Ph.d., Technische Universiteit Eindhoven (2014). (Cited on page 45)
20. Buijs, J., van Dongen, B., van der Aalst, W.M.P.: A genetic algorithm for discovering process trees. In: Proceedings of WCCI 2012 IEEE World Congress on Computational Intelligence, Brisbane, Australia, pp. 925–932 (2012). (Cited on page 45)
21. Burattin, A., Maggi, F.M., Cimitile, M.: Lights, camera, action! business process movies for online process discovery. In: Proceedings of the 3rd International Workshop on Theory and Applications of Process Visualization (TAProViz 2014) (2014). (Cited on page 204)
22. Burattin, A., Maggi, F.M., van der Aalst, W.M.P., Sperduti, A.: Techniques for a posteriori analysis of declarative processes. In: 2012 IEEE 16th International Enterprise Distributed Object Computing Conference, pp. 41–50. IEEE, Beijing, September 2012. (Cited on page 147)
23. Burattin, A., Sperduti, A.: Automatic determination of parameters' values for Heuristics Miner++. In: IEEE Congress on Evolutionary Computation, pp. 1–8. IEEE, Barcelona, Spain, July 2010. (Cited on pages 61 and 97)
24. Burattin, A., Sperduti, A.: Heuristics miner for time intervals. In: European Symposium on Artificial Neural Networks (ESANN), Bruges, Belgium (2010). (Cited on pages 62 and 91)
25. Burattin, A., Sperduti, A.: PLG: a framework for the generation of business process models and their execution logs. In: Muehlen, M., Su, J. (eds.) BPM 2010 Workshops. LNBIP, vol. 66, pp. 214–219. Springer, Heidelberg (2011). (Cited on pages 47 and 164)
26. Burattin, A., Sperduti, A.: Process Log Generator: software documentation (2010). (Cited on page 47)
27. Burattin, A., Sperduti, A., van der Aalst, W.M.P.: Heuristics Miners for Streaming Event Data. ArXiv CoRR, December 2012. (Cited on page 178)
28. Burattin, A., Sperduti, A., van der Aalst, W.M.P.: Control-flow discovery from event streams. In: 2014 IEEE Congress on Evolutionary Computation (CEC), pp. 2420–2427. IEEE, July 2014. (Cited on page 178)
29. Burattin, A., Sperduti, A., Veluscek, M.: Business models enhancement through discovery of roles. In: IEEE Symposium on Computational Intelligence and Data Mining (CIDM). IEEE (2013). (Cited on pages 62 and 120)
30. Burattin, A., Vigo, R.: A framework for semi-automated process instance discovery from decorative attributes. In: IEEE Symposium on Computational Intelligence and Data Mining (CIDM), pp. 176–183. IEEE, Paris, April 2011. (Cited on page 72)
31. Calders, T., Günther, C.W., Pechenizkiy, M., Rozinat, A.: Using minimum description length for process mining. In: Proceedings of the 2009 ACM Symposium on Applied Computing - SAC 2009, pp. 1451–1455. ACM Press, New York (2009). (Cited on pages 103, 104, and 109)
32. Cardoso, J.: Control-flow complexity measurement of processes and Weyuker's properties. Trans. Enformatica Syst. Sci. Eng. **8**, 213–218 (2005). (Cited on page 108)
33. Chang, C.-H., Kayed, M., Girgis, M.R., Shaalan, K.: A survey of web information extraction systems. IEEE Trans. Knowl. Data Eng. **18**(10), 1411–1428 (2006). (Cited on page 23)
34. Chesani, F., Lamma, E., Mello, P., Montali, M., Riguzzi, F., Storari, S.: Exploiting inductive logic programming techniques for declarative process mining. In: Jensen, K., van der Aalst, W.M.P. (eds.) Transactions on Petri Nets and Other Models of Concurrency II. LNCS, vol. 5460, pp. 278–295. Springer, Heidelberg (2009). (Cited on page 147)

35. Cook, J.E.: Process discovery and validation through event-data analysis. Ph.d. thesis, University of Colorado (1996). (Cited on page 35)
36. Cook, J.E., Zhidian, D., Liu, C., Wolf, A.L.: Discovering models of behavior for concurrent workflows. Comput. Ind. **53**(3), 297–319 (2004). (Cited on page 35)
37. Cook, J.E., Wolf, A.L.: Automating process discovery through event-data analysis. In: International Conference on Software Engineering, pp. 73–82. ACM Press (1995). (Cited on page 35)
38. Cook, J.E., Wolf, A.L.: Discovering models of software processes from event-based data. Technical report 3, University of Colorado, November 1996. (Cited on page 35)
39. Cook, J.E., Wolf, A.L.: Balboa: a framework for event-based process data analysis. In: International Conference on the Software Process (1998). (Cited on page 35)
40. Cook, J.E., Wolf, A.L.: Event-based detection of concurrency. ACM SIGSOFT Softw. Eng. Notes **23**(6), 35–45 (1998). (Cited on page 35)
41. Cormen, T.H., Stein, C., Rivest, R.L., Leiserson, C.E.: Introduction to Algorithms, 2nd edn. The MIT Press, Cambridge (2001). (Cited on page 127)
42. Cowie, J., Wilks, Y.: Information extraction. Commun. ACM **39**(1), 80–91 (1996). (Cited on page 23)
43. Davenport, T.H.: Process Innovation: Reengineering Work Through Information Technology. Harvard Business Press, Cambridge (1992). (Cited on page 11)
44. de Medeiros, A.K.A.: Genetic Process Mining. Ph.d. thesis, Technische Universiteit Eindhoven (2006). (Cited on pages 43 and 51)
45. de Medeiros, A.K.A., Günther, C.W.: Process mining: using CPN tools to create test logs for mining algorithms. In: Proceedings of the Sixth Workshop and Tutorial on Practical Use of Coloured Petri Nets and the CPN Tools, pp. 177–190 (2005). (Cited on page 165)
46. de Medeiros, A.K.A., van der Aalst, W.M.P., Weijters, T.A.J.M.M.: Quantifying process equivalence based on observed behavior. Data Knowl. Eng. **64**(1), 55–74 (2008). (Cited on page 138)
47. Deutsch, A., Hull, R., Patrizi, F., Vianu, V.: Automatic verification of data-centric business processes. In: Proceedings of the 12th International Conference on Database Theory - ICDT 2009, pp. 252. ACM Press, New York (2009). (Cited on page 147)
48. Dijkman, R.: Diagnosing differences between business process models. In: Dumas, M., Reichert, M., Shan, M.-C. (eds.) BPM 2008. LNCS, vol. 5240, pp. 261–277. Springer, Heidelberg (2008). (Cited on page 139)
49. Dijkman, R., Dumas, M., van Dongen, B., Käärik, R., Mendling, J.: Similarity of business process models: metrics and evaluation. Inf. Syst. **36**(2), 498–516 (2011). (Cited on page 52)
50. Dumas, M., van der Aalst, W.M.P., ter Hofstede, A.H.M.: Process-Aware Information Systems. Wiley, Hoboken (2005). (Cited on page 42)
51. Ehrig, M., Koschmider, A., Oberweis, A.: Measuring similarity between semantic business process models. In: Proceedings of the Fourth Asia-Pacific Conference on Conceptual Modelling, pp. 71–80 (2007). (Cited on page 138)
52. Ellson, J., Gansner, E.R., Koutsofios, E., North, S.C., Woodhull, G.: Graphviz and Dynagraph Static and Dynamic Graph Drawing Tools. Technical report, AT&T Labs - Research, Florham Park NJ 07932, USA (2004). (Cited on page 172)
53. Elmasri, R., Navathe, S.B.: Fundamentals of Database Systems, 6th edn. Addison-Wesley, Boston (2010). (Cited on pages 75 and 121)
54. Erl, T.: Service-Oriented Architecture: Concepts, Technology and Design. Prentice Hall, Upper Saddle River (2005). (Cited on pages 75 and 121)
55. Ingvaldsen, J.E., Gulla, J.A.: Preprocessing support for large scale process mining of SAP transactions. In: ter Hofstede, A.H.M., Benatallah, B., Paik, H.-Y. (eds.) BPM Workshops 2007. LNCS, vol. 4928, pp. 30–41. Springer, Heidelberg (2008). (Cited on page 82)
56. Ingvaldsen, J.E., Gulla, J.A.: Semantic business process mining of SAP transactions. In: Sun, Z., Wang, M. (eds.) Handbook of Research on Complex Dynamic Process Management: Techniques for Adaptability in Turbulent Environments, Chapter 17, 1st edn, pp. 416–429. Business Science Reference, Hershey (2010). (Cited on page 82)

57. European Commission. Commission Recommendation of 6 May 2003 concerning the definition of micro, small and medium-sized enterprises (2003). (Cited on page 65)

58. Eurostat. European Business: Facts and Figures. European Communities, Luxembourg, Luxembourg (2009). (Cited on page 65)

59. Ferreira, D.R., Gillblad, D.: Discovering process models from unlabelled event logs. In: Dayal, U., Eder, J., Koehler, J., Reijers, H.A. (eds.) BPM 2009. LNCS, vol. 5701, pp. 143–158. Springer, Heidelberg (2009). (Cited on page 81)

60. Ferreira, D., Zacarias, M., Malheiros, M., Ferreira, P.: Approaching process mining with sequence clustering: experiments and findings. In: Alonso, G., Dadam, P., Rosemann, M. (eds.) BPM 2007. LNCS, vol. 4714, pp. 360–374. Springer, Heidelberg (2007). (Cited on page 82)

61. Zaslavsky, A., Krishnaswamy, S.: Mining data streams: a review. ACM SIGMOD Rec. 34(2), 18–26 (2005). (Cited on pages 53 and 54)

62. Ghose, A.K., Koliadis, G.: Auditing business process compliance. In: Krämer, B.J., Lin, K.-J., Narasimhan, P. (eds.) ICSOC 2007. LNCS, vol. 4749, pp. 169–180. Springer, Heidelberg (2007). (Cited on page 46)

63. Giannakopoulou, D., Havelund, K.: Automata-based verification of temporal properties on running programs. In: Proceedings 16th Annual International Conference on Automated Software Engineering (ASE 2001), pp. 412–416. IEEE Computer Society (2001). (Cited on page 147)

64. Goedertier, S., Martens, D., Vanthienen, J., Baesens, B.: Robust process discovery with artificial negative events. J. Mach. Learn. Res. 10, 1305–1340 (2009). (Cited on page 43)

65. Golab, L., Tamer Özsu, M.: Issues in data stream management. ACM SIGMOD Rec. 32(2), 5–14 (2003). (Cited on page 53)

66. Golani, M., Pinter, S.S.: Generating a process model from a process audit log. In: van der Aalst, W.M.P., ter Hofstede, A.H.M., Weske, M. (eds.) BPM 2003. LNCS, vol. 2678, pp. 136–151. Springer, Heidelberg (2003). (Cited on page 41)

67. Golumbic, M.C., Shamir, R.: Complexity and algorithms for reasoning about time: a graph-theoretic approach. J. ACM 40(5), 1108–1133 (1993). (Cited on page 90)

68. Greco, G., Guzzo, A., Pontieri, L.: Mining hierarchies of models: from abstract views to concrete specifications. In: van der Aalst, W.M.P., Benatallah, B., Casati, F., Curbera, F. (eds.) BPM 2005. LNCS, vol. 3649, pp. 32–47. Springer, Heidelberg (2005). (Cited on page 42)

69. Greco, G., Guzzo, A., Pontieri, L., Saccá, D.: Mining expressive process models by clustering workflow traces. In: Dai, H., Srikant, R., Zhang, C. (eds.) PAKDD 2004. LNCS (LNAI), vol. 3056, pp. 52–62. Springer, Heidelberg (2004). (Cited on page 42)

70. Greco, G., Guzzo, A., Pontieri, L., Saccà, D.: Discovering expressive process models by clustering log traces. IEEE Trans. Knowl. Data Eng. 18(8), 1010–1027 (2006). (Cited on page 51)

71. Grünwald, P.: A Tutorial Introduction to the Minimum Description Length Principle. MIT Press, Cambridge (2005). (Cited on pages 102 and 103)

72. Günther, C.W.: Process mining in Flexible Environments. Ph.d. thesis, Technische Universiteit Eindhoven, Eindhoven (2009). (Cited on pages 34, 43, and 61)

73. Günther, C.W., van der Aalst, W.M.P.: A generic import framework for process event logs. In: Eder, J., Dustdar, S. (eds.) BPM Workshops 2006. LNCS, vol. 4103, pp. 81–92. Springer, Heidelberg (2006). (Cited on pages 60 and 116)

74. Günther, C.W., van der Aalst, W.M.P.: Fuzzy mining – adaptive process simplification based on multi-perspective metrics. In: Alonso, G., Dadam, P., Rosemann, M. (eds.) BPM 2007. LNCS, vol. 4714, pp. 328–343. Springer, Heidelberg (2007). (Cited on pages 43 and 44)

75. Günther, C.W., Verbeek, E.H.M.W.: XES Standard Definition (2009). www.xes-standard.org (Cited on pages 60 and 116)

76. Hammer, M., Champy, J.: Reengineering the Corporation: A Manifesto for Business Revolution. Harper Business, New York (1993). (Cited on page 11)

77. Herbst, J.: A machine learning approach to workflow management. In: Lopez de Mantaras, R., Plaza, E. (eds.) ECML 2000. LNCS (LNAI), vol. 1810, pp. 183–194. Springer, Heidelberg (2000). (Cited on page 37)

78. Herbst, J.: Workflow mining with InWoLvE. Comput. Ind. 53(3), 245–264 (2004). (Cited on page 37)

79. Herbst, J., Karagiannis, D.: Integrating machine learning and workflow management to support acquisition and adaptation of workflow models. In: International Workshop on Database and Expert Systems Applications, vol. 9, pp. 745–752. IEEE Computer Society, Los Alamitos, June 1998. (Cited on page 37)

80. Hill, J.B., Sinur, J., Flint, D., Melenovsky, M.J.: Gartner's Position on Business Process Management. Technical report February, Gartner Inc (2006). (Cited on page 19)

81. Hoeffding, W.: Probability inequalities for sums of bounded random variables. J. Am. Stat. Assoc. 58(301), 13–30 (1963). (Cited on page 190)

82. Hwang, S.-Y., Yang, W.-S.: On the discovery of process models from their instances. Decis. Support Syst. 34(1), 41–57 (2002). (Cited on page 38)

83. van der Aalst, W., et al.: Process mining manifesto. In: Daniel, F., Barkaoui, K., Dustdar, S. (eds.) BPM Workshops 2011, Part I. LNBIP, vol. 99, pp. 169–194. Springer, Heidelberg (2012). (Cited on pages 45 and 63)

84. Jensen, K., Kristensen, L.M., Wells, L.: Coloured petri nets and CPN tools for modelling and validation of concurrent systems. Int. J. Softw. Tools Technol. Transf. 9(3–4), 213–254 (2007). (Cited on page 165)

85. Kaiser, K., Miksch, S.: Information Extraction. Technical report May, Vienna University of Technology, Institute of Software Technology and Interactive Systems, Vienna (2005). (Cited on pages 23, and 24)

86. Kalsing, A.C., do Nascimento, G.S., Iochpe, C.,Thom, L.H.: An incremental process mining approach to extract knowledge from legacy systems. In: 2010 14th IEEE International Enterprise Distributed Object Computing Conference, pp. 79–88. IEEE, October 2010. (Cited on page 55)

87. Caskurlu, B.: Model driven engineering. In: Butler, M., Petre, L., Sere, K. (eds.) IFM 2002. LNCS, vol. 2335, pp. 286–298. Springer, Heidelberg (2002). (Cited on page 63)

88. Kindler, E., Rubin, V., Schäfer, W.: Incremental workflow mining based on document versioning information. In: Li, M., Boehm, B., Osterweil, L.J. (eds.) SPW 2005. LNCS, vol. 3840, pp. 287–301. Springer, Heidelberg (2006). (Cited on page 55)

89. Kindler, E., Rubin, V., Schäfer, W.: Incremental workflow mining for process flexibility. In: Proceedings of BPMDS2006, pp. 178–187 (2006). (Cited on page 55)

90. Knuplesch, D., Ly, L.T., Rinderle-Ma, S., Pfeifer, H., Dadam, P.: On enabling data-aware compliance checking of business process models. In: Parsons, J., Saeki, M., Shoval, P., Woo, C., Wand, Y. (eds.) ER 2010. LNCS, vol. 6412, pp. 332–346. Springer, Heidelberg (2010). (Cited on page 147)

91. Ko, R.K.L.: A computer scientist's introductory guide to business process management (BPM). Crossroads 15(4), 11–18 (2009). (Cited on pages 11, 12, and 20)

92. Kopp, O., Martin, D., Wutke, D., Leymann, F.: The difference between graph-based and block-structured business process modelling languages. Enterp. Model. Inf. Syst. 4(1), 3–13 (2009). (Cited on page 39)

93. Kunze, M., Weidlich, M., Weske, M.: Behavioral similarity – a proper metric. In: Rinderle-Ma, S., Toumani, F., Wolf, K. (eds.) BPM 2011. LNCS, vol. 6896, pp. 166–181. Springer, Heidelberg (2011). (Cited on pages 52 and 139)

94. Kupferman, O., Vardi, M.Y.: Vacuity detection in temporal model checking. Int. J. Softw. Tools Technol. Transf. (STTT) 4(2), 224–233 (2003). (Cited on page 149)

95. Leemans, S.J.J., Fahland, D., van der Aalst, W.M.P.: Discovering block-structured process models from event logs - a constructive approach. In: Colom, J.-M., Desel, J. (eds.) PETRI NETS 2013. LNCS, vol. 7927, pp. 311–329. Springer, Heidelberg (2013). (Cited on page 45)

96. de Leoni, M., Maggi, F.M., van der Aalst, W.M.P.: Aligning event logs and declarative process models for conformance checking. In: Barros, A., Gal, A., Kindler, E. (eds.) BPM 2012. LNCS, vol. 7481, pp. 82–97. Springer, Heidelberg (2012). (Cited on page 46)

97. Li, C., Reichert, M., Wombacher, A.: On Measuring Process Model Similarity based on High-level Change Operations. Technical report, Centre for Telematics and Information Technology, University of Twente (2007). (Cited on page 52)

98. Maggi, F.M., Bose, R.P.J.C., van der Aalst, W.M.P.: Efficient discovery of understandable declarative process models from event logs. In: Ralyté, J., Franch, X., Brinkkemper, S., Wrycza, S. (eds.) CAiSE 2012. LNCS, vol. 7328, pp. 270–285. Springer, Heidelberg (2012). (Cited on page 147)

99. Maggi, F.M., Burattin, A., Cimitile, M., Sperduti, A.: Online process discovery to detect concept drifts in LTL-based declarative process models. In: Meersman, R., Panetto, H., Dillon, T., Eder, J., Bellahsene, Z., Ritter, N., De Leenheer, P., Dou, D. (eds.) ODBASE 2013. LNCS, vol. 8185, pp. 94–111. Springer, Heidelberg (2013). (Cited on page 204)

100. Maggi, F.M., Montali, M., Westergaard, M., van der Aalst, W.M.P.: Monitoring business constraints with linear temporal logic: an approach based on colored automata. In: Rinderle-Ma, S., Toumani, F., Wolf, K. (eds.) BPM 2011. LNCS, vol. 6896, pp. 132–147. Springer, Heidelberg (2011). (Cited on page 147)

101. Maggi, F.M., Mooij, A.J., van der Aalst, W.M.P.: User-guided discovery of declarative process models. In: IEEE Symposium on Computational Intelligence and Data Mining (CIDM), pp. 192–199. IEEE, April 2011. (Cited on pages 45, 147, and 149)

102. Manku, G.S., Motwani, R.: Approximate frequency counts over data streams. In: Proceedings of International Conference on Very Large Data Bases, pp. 346–357. Morgan Kaufmann, Hong Kong, China (2002). (Cited on page 187)

103. Manning, C.D., Raghavan, P., Schütze, H.: Introduction to Information Retrieval, vol. 35, 1st edn. Cambridge University Press, Cambridge (2008). (Cited on pages 23, 25, 114 and 130)

104. Manyika, J., Chui, M., Brown, B., Bughin, J., Dobbs, R., Roxburgh, C., Byers, A.H.: Big Data: The Next Frontier for Innovation, Competition, and Productivity. Technical report June, McKinsey Global Institute (2011). (Cited on page 63)

105. Măruşter, L., Weijters, A.J.M.M.T., van der Aalst, W.M.P., van den Bosch, A.: Process mining: discovering direct successors in process logs. In: Lange, S., Satoh, K., Smith, C.H. (eds.) DS 2002. LNCS, vol. 2534, pp. 364–373. Springer, Heidelberg (2002). (Cited on page 141)

106. Mendling, J., van Dongen, B., van der Aalst, W.M.P.: On the degree of behavioral similarity between business process models. In: Workshop on Event-Driven Process Chains, pp. 39–58 (2007). (Cited on page 52)

107. Miller, G.A.: The magical number seven, plus or minus two: some limits on our capacity for processing information. Psychol. Rev. **101**(2), 343–352 (1956). (Cited on page 63)

108. Minor, M., Tartakovski, A., Bergmann, R.: Representation and structure-based similarity assessment for agile workflows. In: Weber, R.O., Richter, M.M. (eds.) ICCBR 2007. LNCS (LNAI), vol. 4626, pp. 224–238. Springer, Heidelberg (2007). (Cited on page 52)

109. Mitchell, T.M.: Machine Learning. McGraw-Hill, New York (1997). (Cited on pages 36, 37, and 43)

110. Muñoz-Gama, J., Carmona, J.: A fresh look at precision in process conformance. In: Hull, R., Mendling, J., Tai, S. (eds.) BPM 2010. LNCS, vol. 6336, pp. 211–226. Springer, Heidelberg (2010). (Cited on pages 51, 192, and 199)

111. Murata, T.: Petri nets: properties, analysis and applications. Proc. IEEE **77**(4), 541–580 (1989). (Cited on page 13)

112. OMG. Business Process Model and Notation (BPMN) - Version 2.0, Beta 1 (2009). (Cited on pages 12, 13 and 15)

113. Ould, M.A.: Business Processes: Modelling and Analysis for Re-engineering and Improvement. Wiley, New York (1995). (Cited on page 12)

114. Papazoglou, M.P., Heuvel, W.-J.: Service oriented architectures: approaches, technologies and research issues. VLDB J. **16**(3), 389–415 (2007). (Cited on page 20)

115. Parrow, J.: Handbook of Process Algebra. Elsevier, Amsterdam (2001). (Cited on page 13)

116. Pérez-castillo, R., Weber, B., Guzmán, I.G.-R., Piattini, M., Pinggera, J. Assessing event correlation in non-process-aware information systems. Softw. Syst. Model., 1–23 (2012). (Cited on page 82)

117. Perrey, R., Lycett, M.: Service-oriented architecture. In: Symposium on Applications and the Internet Workshops, pp. 116–119. IEEE Computer Society (2003). (Cited on page 20)
118. Peterson, J.L.: Petri nets. ACM Comput. Surv. (CSUR) 9(3), 223–252 (1977). (Cited on page 20)
119. Petri, C.A.: Kommumkation mit Automaten. Ph.D. thesis, Institut für Instrumentelle Mathematik, Universität Bonn (1962). (Cited on page 14)
120. Pešić, M.: Constraint-Based Workflow Management Systems: Shifting Control to Users. Ph.d. thesis, Technische Universiteit Eindhoven (2008). (Cited on page 19)
121. Pesic, M., van der Aalst, W.M.P.: A declarative approach for flexible business processes management. In: Eder, J., Dustdar, S. (eds.) BPM Workshops 2006. LNCS, vol. 4103, pp. 169–180. Springer, Heidelberg (2006). (Cited on page 19)
122. Pichler, P., Weber, B., Zugal, S., Pinggera, J., Mendling, J., Reijers, H.A.: Imperative versus declarative process modeling languages: an empirical investigation. In: Daniel, F., Barkaoui, K., Dustdar, S. (eds.) BPM Workshops 2011, Part I. LNBIP, vol. 99, pp. 383–394. Springer, Heidelberg (2012). (Cited on page 18)
123. Pinter, S.S., Golani, M.: Discovering workflow models from activities' lifespans. Comput. Ind. 53(3), 283–296 (2004). (Cited on page 41)
124. Pnueli, A.: The temporal logic of programs. In: 18th Annual Symposium on Foundations of Computer Science (SFCS 1977), pp. 46–57. IEEE, September 1977. (Cited on page 147)
125. Praxiom Research Group Limited. ISO 9000 2005 Plain English Definitions (2009). (Cited on page 12)
126. Rajaraman, A., Ullman, J.D.: Mining of Massive Datasets. Cambridge University Press, Cambridge (2010). (Cited on pages 27 and 144)
127. Vinter Ratzer, A., et al.: CPN tools for editing, simulating, and analysing coloured Petri nets. In: van der Aalst, W.M.P., Best, E. (eds.) ICATPN 2003. LNCS, vol. 2679, pp. 450–462. Springer, Heidelberg (2003). (Cited on page 165)
128. Rozinat, A.: Process Mining: Conformance and Extension. Ph.d., Technische Universiteit Eindhoven (2010). (Cited on page 46)
129. Rozinat, A., de Medeiros, A.K.A., Günther, C.W., Weijters, T.A.J.M.M., van der Aalst, W.M.P.: Towards an evaluation framework for process mining algorithms. BPM Center report BPM-07-06, BPMcenter.org (2007). (Cited on pages 49, and 50)
130. Rozinat, A., van der Aalst, W.M.P.: Decision Mining in Business Processes. Technical report, Business Process Management (BPM) Center (2006). (Cited on page 46)
131. Rozinat, A., van der Aalst, W.M.P.: Decision mining in ProM. In: Dustdar, S., Fiadeiro, J.L., Sheth, A.P. (eds.) BPM 2006. LNCS, vol. 4102, pp. 420–425. Springer, Heidelberg (2006). (Cited on pages 34 and 46)
132. Rozinat, A., van der Aalst, W.M.P.: Conformance checking of processes based on monitoring real behavior. Inf. Syst. 33(1), 64–95 (2008). (Cited on pages 34, 51, and 52)
133. Russell, N., ter Hofstede, A.H.M., van der Aalst, W.M.P., Mulyar, N.: Workflow Control-flow Patterns: A Revised View. BPM Center report BPM-06-22, BPMcenter.org (2006). (Cited on pages 143, 144, and 166)
134. Russell, S., Norvig, P.: Artificial Intelligence: A Modern Approach, 2nd edn. Prentice Hall, Englewood Cliffs (2002). (Cited on page 107)
135. Schimm, G.: Process miner - a tool for mining process schemes from event-based data. In: Flesca, S., Greco, S., Leone, N., Ianni, G. (eds.) JELIA 2002. LNCS (LNAI), vol. 2424, pp. 525–528. Springer, Heidelberg (2002). (Cited on page 39)
136. Schimm, G.: Mining most specific workflow models from event-based data. In: van der Aalst, W.M.P., ter Hofstede, A.H.M., Weske, M. (eds.) BPM 2003. LNCS, vol. 2678, pp. 25–40. Springer, Heidelberg (2003). (Cited on page 39)
137. Schröder, B.: Ordered Sets: An Introduction. Birkhäuser Boston, Boston (2002). (Cited on page 79)
138. Schweikardt, N.: Short-entry on one-pass algorithms. In: Liu, L., Özsu, M.T. (eds.) Encyclopedia of Database Systems, pp. 1948–1949. Springer, Heidelberg (2009). (Cited on page 183)

139. Shannon, C.E.: A mathematical theory of communication. Bell Syst. Tech, J. 27, 379–423, 623–656 (1948). (Cited on page 130)
140. Sharp, A., McDermott, P.: Workflow Modeling: Tools for Process Improvement and Application Development, 2nd edn. Artech House Publishers, Boston (2008). (Cited on page 21)
141. Solé, M., Carmona, J.; Incremental process mining. In: Proceedings of ACSD/Petri Nets Workshops, pp. 175–190 (2010). (Cited on page 55)
142. Song, M., van der Aalst, W.M.P.: Supporting process mining by showing events at a glance. In: Workshop on Information Technologies and Systems (WITS), pp. 139–145 (2007). (Cited on page 116)
143. Song, M., van der Aalst, W.M.P.: Towards comprehensive support for organizational mining. Decis. Support Syst. 46(1), 300–317 (2008). (Cited on page 134)
144. Syropoulos, A.: Mathematics of multisets. In: Calude, C.S., Pun, G., Rozenberg, G., Salomaa, A. (eds.) Multiset Processing. LNCS, vol. 2235, pp. 347–358. Springer, Heidelberg (2001). (Cited on page 122)
145. van der Aalst, W.M.P.: Verification of workflow nets. Appl. Theory Petri Nets 1248, 407–426 (1997). (Cited on pages 14 and 146)
146. van der Aalst, W.M.P.: The application of Petri nets to workflow management. J. Circuits Syst. Comput. 8, 21–66 (1998). (Cited on page 13)
147. van der Aalst, W.M.P.: Business alignment: using process mining as a tool for delta analysis and conformance testing. Requirements Eng. 10(3), 198–211 (2005). (Cited on page 46)
148. van der Aalst, W.M.P.: Process discovery: capturing the invisible. IEEE Comput. Intell. Mag. 5(1), 28–41 (2010). (Cited on page 33)
149. van der Aalst, W., Adriansyah, A., van Dongen, B.: Causal nets: a modeling language tailored towards process discovery. In: Katoen, J.-P., König, B. (eds.) CONCUR 2011. LNCS, vol. 6901, pp. 28–42. Springer, Heidelberg (2011). (Cited on page 183)
150. van der Aalst, W.M.P., Adriansyah, A., van Dongen, B.: Replaying history on process models for conformance checking and performance analysis. Wiley Interdisc. Rev. Data Min. Knowl. Disc. 2(2), 182–192 (2012). (Cited on pages 121 and 192)
151. van der Aalst, W.M.P., de Medeiros, A.K.A., van Dongen, B., Weijters, T.A.J.M.M.: Process Mining: Extending the α-Algorithm to Mine Short Loops. Eindhoven University of Technology, Eindhoven (2004). (Cited on page 40)
152. van der Aalst, W.M.P., de Medeiros, A.K.A., Weijters, T.A.J.M.M.: Using Genetic Algorithms to Mine Process Models: Representation, Operators and Results. BETA Working Paper Series (2004). (Cited on page 107)
153. van der Aalst, W.M.P., de Medeiros, A.K.A., Weijters, T.A.J.M.M.: Genetic process mining. Appl. Theory Petri Nets 3536, 48–69 (2005). (Cited on page 43)
154. van der Aalst, W.M.P., de Medeiros, A.K.A., Weijters, A.J.M.M.T.: Process equivalence: comparing two process models based on observed behavior. In: Dustdar, S., Fiadeiro, J.L., Sheth, A.P. (eds.) BPM 2006. LNCS, vol. 4102, pp. 129–144. Springer, Heidelberg (2006). (Cited on page 138)
155. van der Aalst, W.M.P., Günther, C.W., Rubin, V., Verbeek, E.H.M.W., Kindler, E., van Dongen, B.: Process mining: a two-step approach to balance between underfitting and overfitting. Softw. Syst. Model. 9(1), 87–111 (2008). (Cited on page 101)
156. van der Aalst, W.M.P., Pešić, M., Schonenberg, H.: Declarative workflows: balancing between flexibility and support. Comput. Sci. Res. Dev. 23, 99–113 (2009). (Cited on pages 18 and 147)
157. van der Aalst, W.M.P., Reijers, H.A., Song, M.: Discovering social networks from event logs. Comput. Support. Coop. Work (CSCW) 14(6), 549–593 (2005). (Cited on pages 45 and 135)
158. van der Aalst, W.M.P., Song, M.S.: Mining social networks: uncovering interaction patterns in business processes. In: Desel, J., Pernici, B., Weske, M. (eds.) BPM 2004. LNCS, vol. 3080, pp. 244–260. Springer, Heidelberg (2004). (Cited on page 45)
159. van der Aalst, W.M.P., ter Hofstede, A.H.M.: YAWL: yet another workflow language. inf. syst. 30(4), 245–275 (2005). (Cited on pages 17 and 192)

160. van der Aalst, W.M.P., ter Hofstede, A.H.M., Kiepuszewski, B., Barros, A.P.: Workflow patterns. Distrib. Parallel Databases **14**(1), 5–51 (2003). (Cited on page 17)
161. van der Aalst, W.M.P., ter Hofstede, A.H.M., Weske, M.: Business process management: a survey. In: van der Aalst, W.M.P., ter Hofstede, A.H.M., Weske, M. (eds.) BPM 2003. LNCS, vol. 2678, pp. 1–12. Springer, Heidelberg (2003). (Cited on page 13)
162. van der Aalst, W.M.P., van Dongen, B.F.: Discovering workflow performance models from timed logs. In: Han, Y., Tai, S., Wikarski, D. (eds.) EDCIS 2002. LNCS, vol. 2480, pp. 45–63. Springer, Heidelberg (2002). (Cited on pages 3, 40, and 41)
163. van der Aalst, W.M.P., van Dongen, B.F., Günther, C.W., Mans, R.S., de Medeiros, A.K.A., Rozinat, A., Rubin, V., Song, M., Verbeek, H.M.W.E., Weijters, A.J.M.M.T.: ProM 4.0: comprehensive support for real process analysis. In: Kleijn, J., Yakovlev, A. (eds.) ICATPN 2007. LNCS, vol. 4546, pp. 484–494. Springer, Heidelberg (2007). (Cited on page 60)
164. van der Aalst, W.M.P., van Hee, K.M.: Workflow Management: Models, Methods, and Systems. The MIT Press, Cambridge (2004). (Cited on page 164)
165. van der Aalst, W.M.P., Weijters, T.A.J.M.M.: Rediscovering workflow models from event-based data using little thumb. Integr. Comput.-Aided Eng. **10**(2), 151–162 (2003). (Cited on pages 41 and 54)
166. van der Aalst, W.M.P., Weijters, T.A.J.M.M.: Process mining: a research agenda. Comput. Ind. **53**(3), 231–244 (2004). (Cited on page 3)
167. van der Aalst, W.M.P., Weijters, T.A.J.M.M., de Medeiros, A.K.A.: Process Mining with the Heuristics Miner-algorithm. BETA Working Paper Series, WP 166. Eindhoven University of Technology, Eindhoven (2006). (Cited on pages 12, 41, 51, and 62)
168. van der Aalst, W.M.P., Weijters, T.A.J.M.M., Herbst, J., van Dongen, B., Maruster, L., Schimm, G.: Workflow mining: a survey of issues and approaches. Data Knowl. Eng. **47**(2), 237–267 (2003). (Cited on pages 34 and 39)
169. van der Aalst, W.M.P., Weijters, T.A.J.M.M., Maruster, L.: Workflow Mining: Which processes can be rediscovered? Technical report, Eindhoven University of Technology, Eindhoven (2002). (Cited on page 40)
170. van der Aalst, W.M.P.: Process Mining: Discovery, Conformance and Enhancement of Business Processes. Springer, Heidelberg (2011). (Cited on pages 141 and 178)
171. van der Werf, J.M.E.M., van Dongen, B., van Hee, K.M., Hurkens, C.A.J., Serebrenik, A.: Process discovery using integer linear programming. Computer Science report (08–04), Eindhoven University of Technology, Eindhoven, The Netherlands (2008). (Cited on page 45)
172. van Dongen, B., Busi, N., Pinna, G.M., van der Aalst W.M.P.: An iterative algorithm for applying the theory of regions in process mining. In: Workshop on Formal Aspects of Business Processes and Web Services (2007). (Cited on page 45)
173. van Dongen, B., de Medeiros, A.K.A., Verbeek, E.H.M.W., Weijters, T.A.J.M.M., van der Aalst, W.M.P.: The ProM framework: a new era in process mining tool support. Appl. Theory of Petri Nets **3536**, 444–454 (2005). (Cited on pages 60 and 115)
174. van Dongen, B.F., Dijkman, R., Mendling, J.: Measuring similarity between business process models. In: Bellahsène, Z., Léonard, M. (eds.) CAiSE 2008. LNCS, vol. 5074, pp. 450–464. Springer, Heidelberg (2008). (Cited on page 139)
175. van Dongen, B.F., van der Aalst, W.M.P.: Multi-phase process mining: building instance graphs. In: Atzeni, P., Chu, W., Lu, H., Zhou, S., Ling, T.-W. (eds.) ER 2004. LNCS, vol. 3288, pp. 362–376. Springer, Heidelberg (2004). (Cited on page 42)
176. van Dongen, B., van der Aalst, W.M.P.: Multi-phase process mining: aggregating instance graphs into EPCs and Petri nets. In: PNCWB 2005 Workshop, pp. 35–58 (2005). (Cited on page 42)
177. van Hee, K.M., Liu, Z.: Generating benchmarks by random stepwise refinement of Petri nets. In: Proceedings of Workshop APNOC/SUMo (2010). (Cited on page 164)
178. van Leeuwen, M., Siebes, A.: STREAMKRIMP: detecting change in data streams. In: Daelemans, W., Goethals, B., Morik, K. (eds.) ECML PKDD 2008, Part I. LNCS (LNAI), vol. 5211, pp. 672–687. Springer, Heidelberg (2008). (Cited on page 178)

179. Verbeek, E.H.M.W., Buijs, J., van Dongen, B., van der Aalst, W.M.P.: ProM 6: the process mining toolkit. In: BPM 2010 Demo, pp. 34–39 (2010). (Cited on pages 60 and 115)

180. Walicki, M., Ferreira, D.R.: Mining sequences for patterns with non-repeating symbols. In: IEEE Congress on Evolutionary Computation 2010, Barcelona, Spain, pp. 3269–3276 (2010). (Cited on page 82)

181. Wang, J., He, T., Wen, L., Wu, N., ter Hofstede, A.H.M., Su, J.: A behavioral similarity measure between labeled petri nets based on principal transition sequences. In: Meersman, R., Dillon, T.S., Herrero, P. (eds.) OTM 2010. LNCS, vol. 6426, pp. 394–401. Springer, Heidelberg (2010). (Cited on page 139)

182. Weidlich, M., Mendling, J., Weske, M.: Efficient consistency measurement based on behavioral profiles of process models. IEEE Trans. Softw. Eng. 37(3), 410–429 (2011). (Cited on pages 139 and 142)

183. Wen, L., Wang, J., van der Aalst, W.M.P., Huang, B., Sun, J.-G.: A novel approach for process mining based on event types. J. Intel. Inf. Syst. 32(2), 163–190 (2008). (Cited on page 40)

184. Widmer, G., Kubat, M.: Learning in the presence of concept drift and hidden contexts. Mach. Learn. 23(1), 69–101 (1996). (Cited on pages 54 and 178)

185. Wing, J.M.: FAQ on Pi-Calculus, December 2002. (Cited on page 13)

186. Zha, H., Wang, J., Wen, L., Wang, C., Sun, J.-G.: A workflow net similarity measure based on transition adjacency relations. Comput. Ind. 61(5), 463–471 (2010). (Cited on pages 139, 140, 141, 142, 146)

187. Zugal, S., Pinggera, J., Weber, B.: The impact of testcases on the maintainability of declarative process models. In: Halpin, T., Nurcan, S., Krogstie, J., Soffer, P., Proper, E., Schmidt, R., Bider, I. (eds.) BPMDS 2011 and EMMSAD 2011. LNBIP, vol. 81, pp. 163–177. Springer, Heidelberg (2011). (Cited on page 18)

188. Zurawski, R.: Petri nets and industrial applications: a tutorial. IEEE Trans. Industr. Electron. 41(6), 567–583 (1994). (Cited on page 13)

Printed in the United States
By Bookmasters